# Model Based Inference in the
# Life Sciences: A Primer on Evidence

David R. Anderson

# Model Based Inference in the Life Sciences: A Primer on Evidence

 Springer

David R. Anderson
Colorado Cooperative Fish & Wildlife Research Unit
Department of Fish, Wildlife and Conservation Biology
Colorado State University
Fort Collins, CO 80523
USA
anderson@cnr.colostate.edu

ISBN: 978-0-387-74073-7          e-ISBN: 978-0-387-74075-1

Library of Congress Control Number: 2007934534

The photograph on the cover is of Auguste Rodin's bronze sculpture, The Thinker. The statue has become an icon representing intellectual activity and hence, reflects a focus of this textbook. Deep thought is required to hypothesize an array of plausible science hypotheses. Methods are now available to provide a strength of evidence for members of this array and these are simple to compute and understand. It is this hard thinking that is so vital to empirical science. Photo was courtesy of Dick Johnsen.

Printed on acid-free paper.

9 8 7 6 5 4 3 2 (Corrected at 2nd printing 2010)

springer.com

*To my daughters, Tamara E. and Adrienne M. Anderson*

# Preface

This book is about the "information-theoretic" approaches to rigorous inference based on Kullback–Leibler information. My objective in writing this book is to provide an introduction to making rigorous statistical inferences from data and models about hypotheses in the life sciences. The goal of this primer is to explain the information-theoretic approaches with a focus on application. I stress science philosophy as much as statistical theory and I wade into some ideas from information theory because it is so interesting. The book is about hypothesizing science alternatives and providing quantitative evidence for these.

In 1973 Hirotugu Akaike made a world class discovery when he found a linkage between K–L information and statistical theory through the maximized log-likelihood function. Statistical theory developed since the mid-1970s allows science hypotheses, represented by mathematical models, to be ranked from best to worst. In addition, the discrete probability of each hypothesis $i$, given the data, can be easily computed. These can be viewed as Bayesian posterior model probabilities and are quite important in making inferences about the science hypotheses of interest. The likelihood of each hypothesis, given the data, and evidence ratios between hypotheses $i$ and $j$ are also available, and easy to interpret. All of these new features are simple to compute and understand and go far beyond traditional methods. While many of the examples are biological, I hope students and scientists in other fields (e.g., social sciences, medicine, economics, and many other disciplines) can learn from this primer. Several examples are ecological as that has been my interest; however, the specific examples used are far less important than the science context and trying to understand new approaches; I could not include an example from all of the many subdisciplines.

Tosio Kitagawa (1986) noted that the information-theoretic methods are

*"... a challenge to conventional statistics as well as a proposal for a new approach to statistical analysis. The reader may find some aspects of the approach controversial insofar as they imply a criticism of conventional mathematical statistics, such as the use of statistical tests, individual sampling distribution theory, and statistical tables."*

I find that some people are still struggling with these new approaches 20 years later. Perhaps this reticence is healthy for science as new ideas must be carefully evaluated and scrutinized; however, we must not let "progress ride on a hearse" either.

I have tried to write this as a science textbook; in a sense it is a companion to the books I have written on this subject with Ken Burnham in 1998 and 2002. Those books contain statistical theory, derivations, proofs, some comparisons with other approaches, and were written at a more advanced level. The present primer tries to be well above a "cookbook" but well below a highly technical treatment; this is a book largely for people new to these effective approaches to empirical science. The book provides a consistent strategy (the concepts of evidence and evolving hypothesis sets) for rapid learning and a way of thinking about science and discovery; a road map of sorts. I provide several examples and many insights on modeling; however, I must say clearly that this is not a primer on modeling.

In the first 4 chapters I cover some material to motivate thinking about plausible science hypotheses (the most important issue), data, information, K–L information, and various measures of evidence and support for members of a set of science hypotheses and their corresponding models. Several examples continue through the chapters as new developments are introduced. At this point, the basics of model based inference under the "information-theoretic" approach will have been laid out. But then, like many good novels – there is a twist. Instead of trying to identify the best science hypothesis (and its model) from the set of hypotheses, I refocus on making formal inference based on all the models – "multimodel inference." In many cases it is desirable to make predictions from all the hypotheses in an *a priori* set – one facet of multimodel inference. These procedures allow model averaging and unconditional measures of precision. Those people thinking this jump will surely be difficult will be pleasantly surprised. The main approaches to multimodel inference under this approach can be understood in 1–2 h of lecture and discussion – they are relatively simple but effective. I hope readers will conceptualize their empirical work in science as multimodel inference. This mental image will help focus on the importance of deriving a set of good, plausible science hypotheses (the hard thinking), gathering quality data, and using modern methods to provide quantitative evidence for each of the science hypotheses of interest.

I want to be quick to say that there are other valid approaches to making inferences from empirical data and I make no effort to deny these. There are general theories related to cross validation, nonparametric statistics, bootstrapping, and

Bayesian approaches to mention only a few. In addition, there are a number of new theories for model selection for linear models; I have omitted reference to these special cases but admit that, with further development, they may someday have wider application. Of the four general theories I noted, only the Bayesian approaches have the breadth and depth of those based on information theory. All have their strengths and I encourage some understanding of these approaches. I will make passing reference to some of these alternatives. I am pro-Bayesian and am interested in areas of commonality between the information-theoretic methods and Bayesian methods. Frequentists and Bayesians have waged a long and protracted philosophical war; I do not want to see the information-theoretic approaches join the conflict.

I consider the various null hypothesis testing approaches to be only of historical interest at this stage (2007), except perhaps in the analysis of data from strict experiments where the design stipulates a single model (i.e., design based inference). In general I think scientists serious about their work must move beyond testing sterile null hypotheses to modern methods and the substantial advantages they provide. I offer several comparisons.

This primer is written to be useful for seniors in excellent undergraduate science programs at top universities. Perhaps more realistically, the book is aimed at graduate students, post-doctoral fellows, as well as established scientists in academia, government agencies, and various science institutes. A basic statistical background is essential to easily understand the material in this book: sampling theory, simple experimental designs, measures of variability and covariability (e.g., sampling variances and covariances, standard errors, coefficients of variation, various approaches to confidence intervals, and sampling correlations), "regression" (e.g., $\beta_i$ as partial regression coefficients, $R^2$, residual variance $\sigma^2$, residual sum of squares RSS), and goodness-of-fit concepts.

Ideally, the reader would have had some introduction to Fisher's likelihood approaches (e.g., maximum likelihood estimates, profile likelihood intervals). It is hard to understand why there is so much emphasis on least squares approaches even in graduate courses for nonstatistics majors as this narrow approach comes at the expense of the much more general and useful likelihood methods. In addition, likelihood is foundational to the Bayesian approaches. Readers with the required background will find the quantitative issues easy; it is the deeper conceptual issues that will challenge nearly everyone (e.g., model selection bias). This is the fun and rewarding part of science – thinking hard. Readers lacking exposure to null hypothesis testing will find the material here easier to understand than their counterparts. Still, readers should expect to have to reread some material and contemplate the examples given to chase a full understanding of the material.

A *Remarks* section is found near the end of most chapters and some people will find these unordered comments interesting; however, I suggest this material might best be saved for a second reading. This material includes historical comments, technical notes, and other tangential issues that I thought might interest many readers. In a sense, the *Remarks* are a grouping of what would otherwise be "footnotes," which I often find interesting, but sometimes distracting from the main points. Most chapters end with some nontraditional exercises. Comments on answers to some of the exercises can be found at www.springer.com/978-0-387-74073-7 Each chapter includes a photo and short biography of people who have made major contributions to this literature. I think it is important to recognize and learn from people who came before us and made substantial contributions to science.

This is not an introductory text as I assume a basic knowledge of statistics, the ability to conceptualize science hypotheses ($H_i$), represent these by mathematical models ($g_i$), obtain estimates of model parameters ($\theta$), their sampling covariance matrix ($\Sigma$), goodness-of-fit tests, and residual analysis. Given this backdrop, new and deeper questions can be asked and answers quantified effectively and simply. I believe this material is fun and exciting if you are a scientist who is serious about advanced work on problems where there are substantial stochasticities and complexities. The material is very broad, but I say less about models for multivariate responses and random effects (as opposed to so-called fixed effects) models.

Many people in the life sciences leave graduate programs with little or no exposure to quantitative thinking and methods and this is an increasingly serious issue, limiting both their contributions to science and their career growth. Many PhD-level people lack any working knowledge of calculus, statistics, matrix algebra, computer programming, numerical methods, and modeling. Some people think that is why they are in biology – "because then I don't have to learn that quantitative stuff." I can certainly understand the sentiment; however, there are ample reasons to reconsider, even later in life. Quantification becomes essential in real world problems as a science matures in a given discipline.

In a sense, undergraduate students are taught a small fraction of material that is *already known* in their field and associated disciplines. People need this background information. Graduate work is quite different (or should be), as students are taught effective philosophies and methods to help them learn how to understand things *new* to their field of science. First one wants to know the current "edge" of knowledge on some issue. Second, one wants to push that edge further as new things are learned from the science process. These are things that cannot be found in a book or on the Internet; the discovery of *new* things – this is what science is all about. Good undergraduate programs try to blur the line between these extremes and this is healthy for science. Graduate programs try to help students shift gears into considering methodologies and philosophies for rapid learning of new things; things that no one has discovered (yet). These are often the harder, more complex issues as our predecessors have solved the easier problems. The information-theoretic approaches represent an effective science strategy and allow one to shift into 6th or even 7th gear and that is what makes learning this material both important and fun.

I wanted to write a short book and try to make some main points that I think are important to people coming to these subjects for the first time. This is a book about doing empirical science. I do not expect everyone to agree with every philosophical or analytical aspect. Few of the ideas are originally mine as I have taken from thoughts and results from many others as I try to synthesize the more fundamental issues for the reader. This synthesis comes from about 40 years of experience, studying the work of others and trying to form a coherent philosophy about an effective way to do empirical science. I hope people will take what they find useful and be willing to forge ahead in areas where they have found better approaches. I intend to remain interested in this broad subject and will always enjoy hearing comments from colleagues, many of whom I have not yet met.

I want to fully acknowledge my closest friend and colleague over the last 34 years, Ken Burnham. I (reluctantly) wrote this text alone as I am trusting Ken to complete his book on experimental design. Ken has had an powerful influence on my thinking about science philosophy, statistics, information theory, and model based inference. Several other people helped in various ways and I am proud to acknowledge Peter Caley and Jim Hone for their help with my use of their ferret data and Lianne Ball and Paul Doherty for their help with the Palm Springs ground squirrel example. I benefited greatly from extensive review comments offered by Peter Beerli, Barry Grand, Benedikt Schmidt, and Bill Thompson. I also want to thank Bill Gould, Paul Lukacs, Dave Otis, and Eric Stolen for their advice and encouragement. The photo of Thomas Chamberlin was provided by the Edgar Fahs Smith collection at the University of Pennsylvannia. John Kimmel at Springer was both patient and encouraging as he helped me through the writing and publishing process.

Fort Collins, CO                                          David R. Anderson
                                                              June, 2007

# Contents

Preface ......................................................................................... vii

About the Author ......................................................................... xvii

Glossary ...................................................................................... xix

1.  **Introduction: Science Hypotheses and Science Philosophy** ............. **1**

    1.1    Some Science Background ............................................... 1
    1.2    Multiple Working Hypotheses .......................................... 3
    1.3    Bovine TB Transmission in Ferrets ................................. 4
    1.4    Approaches to Scientific Investigations ........................... 6
           1.4.1   Experimental Studies ........................................ 7
           1.4.2   Descriptive Studies .......................................... 8
           1.4.3   Confirmatory Studies ....................................... 8
    1.5    Science Hypothesis Set Evolves ..................................... 10
    1.6    Null Hypothesis Testing ................................................. 11
    1.7    Evidence and Inferences ................................................. 12
    1.8    Hardening of Portland Cement ....................................... 13
    1.9    What Does Science Try to Provide? ................................ 14
    1.10   Remarks ...................................................................... 15
    1.11   Exercises ...................................................................... 17

2.  **Data and Models** ................................................................. **19**

    2.1    Data ............................................................................. 19
           2.1.1   Hardening of Portland Cement Data ................... 22
           2.1.2   Bovine TB Transmission in Ferrets .................... 23
           2.1.3   What Constitutes a "Data Set"? ........................ 24

2.2   Models ....................................................................................... 25
    2.2.1   True Models (An Oxymoron) ........................................ 27
    2.2.2   The Concept of Model Parameters ............................... 28
    2.2.3   Parameter Estimation ................................................... 29
    2.2.4   Principle of Parsimony ................................................ 30
    2.2.5   Tapering Effect Sizes ................................................... 33
2.3   Case Studies ............................................................................. 33
    2.3.1   Models of Hardening of Portland Cement Data .......... 33
    2.3.2   Models of Bovine TB Transmission in Ferrets ........... 35
2.4   Additional Examples of Modeling ......................................... 36
    2.4.1   Modeling Beak Lengths ............................................... 37
    2.4.2   Modeling Dose Response in Flour Beetles .................. 41
    2.4.3   Modeling Enzyme Kinetics .......................................... 44
2.5   Data Dredging .......................................................................... 45
2.6   The Effect of a Flood on European
      Dippers: Modeling Contrasts ................................................. 46
    2.6.1   Traditional Null Hypothesis Testing ........................... 46
    2.6.2   Information-Theoretic Approach ................................. 47
2.7   Remarks .................................................................................... 48
2.8   Exercises ................................................................................... 49

3.   Information Theory and Entropy ....................................................... 51

3.1   Kullback–Leibler Information ................................................. 52
3.2   Linking Information Theory to Statistical Theory ................. 54
3.3   Akaike's Information Criterion ............................................... 55
    3.3.1   The Bias Correction Term ........................................... 57
    3.3.2   Why Multiply by −2? ................................................... 57
    3.3.3   Parsimony is Achieved as a by-Product ...................... 58
    3.3.4   Simple vs. Complex Models ........................................ 59
    3.3.5   AIC Scale ..................................................................... 60
3.4   A Second-Order Bias Correction: AICc .................................. 60
3.5   Regression Analysis ................................................................. 61
3.6   Additional Important Points .................................................... 62
    3.6.1   Differences Among AICc Values ................................. 62
    3.6.2   Nested vs. Nonnested Models ..................................... 63
    3.6.3   Data and Response Variable Must Remain Fixed ....... 63
    3.6.4   AICc is not a "Test" .................................................... 64
    3.6.5   Data Dredging Using AICc .......................................... 64
    3.6.6   Keep all the Model Terms ........................................... 64
    3.6.7   Missing Data ................................................................ 65
    3.6.8   The "Pretending Variable" ........................................... 65
3.7   Cement Hardening Data ........................................................... 66
    3.7.1   Interpreting AICc Values ............................................. 66
    3.7.2   What if all the Models are Bad? .................................. 67
    3.7.3   Prediction from the Best Model ................................... 68

3.8    Ranking the Models of Bovine Tuberculosis in Ferrets ........    69
3.9    Other Important Issues .........................................................    70
        3.9.1    Takeuchi's Information Criterion............................    70
        3.9.2    Problems When Evaluating Too Many
                 Candidate Models .....................................................    71
        3.9.3    The Parameter Count K and Parameters
                 that Cannot be Uniquely Estimated .........................    71
        3.9.4    Cross Validation and AICc .......................................    72
        3.9.5    Science Advances as the Hypothesis
                 Set Evolves................................................................    72
3.10   Summary...............................................................................    73
3.11   Remarks................................................................................    74
3.12   Exercises...............................................................................    80

4.   **Quantifying the Evidence About Science Hypotheses**.....................    **83**
4.1    $\Delta_i$ Values and Ranking.....................................................    84
4.2    Model Likelihoods................................................................    86
4.3    Model Probabilities .............................................................    87
4.4    Evidence Ratios ...................................................................    89
4.5    Hardening of Portland Cement............................................    91
4.6    Bovine Tuberculosis in Ferrets............................................    93
4.7    Return to Flather's Models and $R^2$....................................    94
4.8    The Effect of a Flood on European Dippers.........................    95
4.9    More about Evidence and Inference.....................................    98
4.10   Summary...............................................................................    100
4.11   Remarks................................................................................    101
4.12   Exercises...............................................................................    103

5.   **Multimodel Inference** ...........................................................    **105**
5.1    Model Averaging..................................................................    106
        5.1.1    Model Averaging for Prediction................................    107
        5.1.2    Model Averaging Parameter
                 Estimates Across Models .........................................    108
5.2    Unconditional Variances......................................................    110
        5.2.1    Examples Using the Cement Hardening Data.............    112
        5.2.2    Averaging Detection Probability Parameters
                 in Occupancy Models ...............................................    115
5.3    Relative Importance of Predictor Variables...........................    118
        5.3.1    Rationale for Ranking the Relative Importance
                 of Predictor Variables .............................................    119
        5.3.2    An Example Using the Cement Hardening Data ........    119
5.4    Confidence Sets on Models...................................................    121
5.5    Summary...............................................................................    122
5.6    Remarks................................................................................    122
5.7    Exercises...............................................................................    124

6.  **Advanced Topics** ....................................................... **125**

    6.1   Overdispersed Count Data............................................ 126
          6.1.1   Lack of Independence ..................................... 126
          6.1.2   Parameter Heterogeneity ................................. 126
          6.1.3   Estimation of a Variance Inflation Factor.............. 127
          6.1.4   Coping with Overdispersion in Count Data .............. 127
          6.1.5   Overdispersion in Data on Elephant Seals .............. 128
    6.2   Model Selection Bias................................................ 129
          6.2.1   Understanding the Issue ................................. 129
          6.2.2   A Solution to the Problem of Model
                  Selection Bias........................................... 130
    6.3   Multivariate AICc.................................................. 133
    6.4   Model Redundancy.................................................. 133
    6.5   Model Selection in Random Effects Models...................... 134
    6.6   Use in Conflict Resolution ...................................... 135
          6.6.1   Analogy with the Flip of a Coin........................ 136
          6.6.2   Conflict Resolution Protocol........................... 137
          6.6.3   A Hypothetical Example: Hen Clam
                  Experiments............................................. 138
    6.7   Remarks............................................................ 140

7.  **Summary**............................................................... **141**

    7.1   The Science Question .............................................. 142
    7.2   Collection of Relevant Data ...................................... 143
    7.3   Mathematical Models .............................................. 143
    7.4   Data Analysis...................................................... 144
    7.5   Information and Entropy ........................................... 144
    7.6   Quantitative Measures of Evidence................................ 144
    7.7   Inferences ........................................................ 145
    7.8   *Post Hoc* Issues................................................. 146
    7.9   Final Comment ..................................................... 146

**Appendices**................................................................ **147**

    Appendix A: Likelihood Theory ........................................... 147
    Appendix B: Expected Values............................................. 155
    Appendix C: Null Hypothesis Testing..................................... 157
    Appendix D: Bayesian Approaches......................................... 158
    Appendix E: The Bayesian Information Criterion.......................... 160
    Appendix F: Common Misuses and Misinterpretations...................... 162

**References**............................................................... **167**

**Index**.................................................................... **181**

# About the Author

David R. Anderson received B.S. and M.S. degrees in wildlife biology from Colorado State University and a Ph.D. in theoretical ecology from the University of Maryland. He spent 9 years as a research biologist at the Patuxent Wildlife Research Center in Maryland and 9 years as Leader of the Utah Cooperative Wildlife Research Unit and professor in the Wildlife Science Department at Utah State University. He was a Senior Scientist with the Biological Resources Division within the U.S. Geological Survey, Leader of the Colorado Cooperative Fisheries and Wildlife Research Unit, and a professor in the Department of Fishery and Wildlife Biology at Colorado State University until his retirement in 2003. He has been at Colorado State University since 1984 where he holds an emeritus professorship. He is president of the Applied Information Company in Fort Collins, Colorado. He has published 18 books and monographs, 107 papers in peer reviewed national and international journals, and 45 book chapters, government scientific reports, conference proceedings, and transactions. He is the recipient of numerous professional awards for scientific and academic contributions, including the 2004 Aldo Leopold Memorial Award and Medal.

# Glossary

## Terms

| | |
|---|---|
| Akaike weight | The probability that model $i$ is the actual (fitted) K–L best model in the set |
| Asymptotic | A result or procedure where sample size goes to infinity as a limit |
| Bias | (Of an estimator) Bias = $E(\hat{\theta}) - \theta$. |
| Deductive inference | Reasoning from the general to the particular. Central in logic |
| Deviance | A fundamental term in likelihood theory. In this book we can usually get by with deviance = $-2 \log \mathcal{L}$, that is, negative 2 times the value of the log-likelihood at its maximum point |
| Effect size | A general term to reflect a measure of the magnitude of some parameter. In a simple experiment, the effect size is often the difference in treatment vs. control means, $\mu_c - \mu_t$. In regression and other models, the effect size is just the (regression) parameter, $\beta_j$. In some survival studies, the effect size is defined as the ratio of treatment and control survival probabilities, $\varphi_t / \varphi_c$ |
| Entropy | A measure of disorder or randomness. A highly technical issue as there is more than one form. A short introduction is given in the *Remarks* section in Chap. 3 |
| Estimate | The computed value of an estimator, given a particular set of sample data (e.g., $\hat{\theta} = 9.8$) |
| Estimator | A function of the sample data that is used to estimate some parameter. A simple example is $\hat{p} = y / n$ for a binomial proportion. An estimator is a random variable and denoted by a "hat" (e.g., $\hat{p}$ or $\hat{\theta}$). Some estimators do |

|                      |                                                                 |
|----------------------|-----------------------------------------------------------------|
|                      | not have a simple "closed form" and rely on numerical methods to compute their numerical value |
| Evidence ratio       | A ratio of the model probabilities for models $i$ and $j$ in the set, $E_{i,j}$. Used as a quantitative measure of the strength of evidence for any two hypotheses $i$ and $j$ |
| Global model         | Usually the most highly dimensioned model in the set; used primarily for goodness of fit assessment. At least some models in the set are often nested within the global model |
| iid                  | Abbreviation for "independent and identically distributed" |
| Inductive inference  | Reasoning from a sample to the population from which the sample was drawn. Central to statistical inference and fundamental to empirical science |
| Likelihood           | A relative value useful in comparing entities. Not a probability as likelihoods do not sum or integrate to 1. Likelihoods are 0 or positive. For example, one can compute the likelihood of various values of $p$, given the data ($n$ and $y$) and the binomial model. A single likelihood value is not useful; at least one more value is needed as likelihood values are relative (comparative) to some reference value. Appendix A |
| Log-likelihood       | The natural logarithm of the likelihood function and fundamental in both statistical and information theory |
| Mean squared error   | A measure of performance or accuracy, often in prediction, and defined as the sum of squared bias + variance |
| Model probability    | The discrete probability of model $i$ being the actual best model in terms of K–L information |
| Negentropy           | The negative of entropy, also equal to K–L information |
| Nested models        | A model that is a special case of another model is said to be "nested." A linear model $E(Y) = \beta_0 + \beta_1(x)$ is nested within the quadratic model $E(Y) = \beta_0 + \beta_1(x) + \beta_2(x^2)$ |
| Occam's razor        | Taken from thirteenth-century English monk is the well worn idea of the importance of simplicity. The "razor" is the concept "shave away all that is unnecessary" |
| Parsimony            | Classically, this concept is a bias versus variance trade-off. It implies a balancing between the evils of over-fitting and under-fitting. This term should not mean just a "smaller" model as it is sometimes used. Instead, parsimony refers to some trade-off between too few and too many parameters, given a particular sample size. Closely related to Occam's razor |

Precision            A property of an estimator related to the amount of variation among estimates from repeated samples. Precision is measured by the sampling variance, standard error, coefficient of variation, and various types of confidence interval. Precision and information are closely related

Predictive mean squared error        Conceptually the expected value of the variance + squared bias. Practically, this can be estimated as $E[(\hat{Y}_i) - E(Y_i)]^2$, where $\hat{Y}_i$ is the predicted value from the $i$th sample

Pretending variable        Slang for the case where a model containing an unrelated variable enters the model set with a $\Delta$ value of about 2 and is judged as being a "good" model; however, the deviance was not changed. Thus the variable is "pretending" to be important by being in a "good" model, but since the fit was not improved, the variable must be recognized as unimportant. Further evidence of this can be gleaned from examination of the confidence interval for the associated parameter estimate

Probability          Many people consider probabilities to be only long-term frequencies; others (e.g., Bayesians) have the expanded view that probabilities can convey a quantification of belief. In either case, they are nonnegative quantities and sum or integrate to 1 and range between 0 and 1, inclusive

## Symbols

AIC          Akaike's Information Criterion, $= -2\log(\mathcal{L}(\theta|x)) + 2K$ or just $-2\log(\mathcal{L}) + 2K$ in shorthand notation

$\text{AIC}_{\min}$        The estimate of expected K–L information for the best model in the set, given the data. For example, given the model set $(g_1, g_2,..., g_R)$ and the data $x$, if the information criterion is minimized for model $g_6$, then min $= 6$, signifying that $\text{AIC}_6$ is the minimum over $\text{AIC}_1$, $\text{AIC}_2$,..., $\text{AIC}_R$. The minimum AIC is a random variable over samples. This notation, indicating the index number in $\{1, 2,..., R\}$ that minimizes expected K–L information, also applies to AICc, QAICc, and TIC

AICc          A second-order AIC, useful when sample size is small in relation to the number of model parameters to be estimated $(K)$. $\text{AICc} = -2\log(\mathcal{L}(\theta|x) + 2K + 2(K(K+1))/(n-K-1)$

$\beta_j$          Standard notation for a (partial) regression coefficient ("slopes" relating to the $j$th predictor variable)

| | |
|---|---|
| BIC | Bayesian Information Criterion (also termed SIC in some literature for Schwarz's information criterion) |
| $cov(\hat{\theta}_i, \hat{\theta}_j)$ | The sampling covariance of two estimators $\hat{\theta}_i$ and $\hat{\theta}_j$, respectively. This is a measure of codependence and reflects the fact that both estimates, $i$ and $j$, come from the same data set and, therefore, might be related (dependent) |
| $c$ | A simple variance inflation factor used in quasi-likelihood methods where there is overdispersion of count data (e.g., extrabinomial variation). $c \equiv 1$ under independence |
| $\Delta_i$ | AIC differences, relative to the smallest AIC value in the model set. The best model has $\Delta_i \equiv 0$. Formally, $\Delta_i = AIC_i - AIC_{min}$). These values are estimates of the expected K–L information (or distance) between the best (selected) model and the $i$th model. These differences apply to AIC, AICc, QAICc, and TIC |
| $e_i$ | The $i$th residual in regression analysis, $y_i - \hat{y}_i$ |
| $E(\hat{\theta})$ | An operator meaning to take the statistical expectation of the estimator $\hat{\theta}$. Roughly an average of the parameter estimates taken over an infinite number of realizations from the stochastic process that generated the data for a fixed sample size (Appendix B) |
| $E_{i,j}$ | The evidence ratio; the relative likelihood of hypothesis $i$ vs. hypothesis $j$ or, equivalently, model $i$ versus model $j$. A formal measure of the strength of evidence of any two science hypotheses $i$ and $j$ in the candidate set |
| $f(x)$ | Used to denote hypothetical "truth" or "full reality," the process that produces multivariate data, $x$. This conceptual or hypothetical "probability distribution" is considered to be infinite dimensional (i.e., an infinite number of "entities," not necessarily what we term "parameters") |
| $g_i(x)$ | Used to denote the model representing science hypotheses $i$. These models are a function of the data ($x$), thus the notation $g_i(x)$. The set of $R$ candidate models is represented simply as $g_1, g_2,..., g_R$ |
| GOF | Goodness-of-fit test or statistic |
| $H_i$ | The $i$th science hypothesis |
| $H_o$ | The "null" hypothesis, the hypothesis tested in null hypothesis testing |
| $H_a$ | The "alternative" hypothesis associated with null hypothesis testing |

| | |
|---|---|
| $K$ | The number of estimable parameters in an approximating model. Some parameters are confounded with another in some models and are then not "identifiable." In such cases, the parameter count ($K$) should add 1 parameter for the confounded pair (not 2) |
| K–L | Kullback–Leibler information (or distance, discrepancy, number) |
| LS | Least squares method of estimation ("regression") |
| $\mathcal{L}(\theta\|x)$ | Likelihood function of the model parameters, given the data $x$ |
| $\mathcal{L}(\theta\|x, g_i)$ | Extended notation to denote the fact that the likelihood function always assumes the data and the specific model $g_i$ are *given* |
| $\log(\cdot)$ | The natural logarithm ($\log_e$). All logarithms in this book are natural (Naperian) logarithms |
| $\log(\mathcal{L})$ | Shorthand notation for the log-likelihood function |
| $\log(\mathcal{L}(\theta\|x, g_i))$ | Extended notation to denote the fact that the log-likelihood function always assumes the data and the specific model are given |
| $\text{logit}(\theta)$ | The logit transform: $\text{logit}(\theta) = \log(\theta/(1-\theta))$, where $0 < \theta < 1$ |
| $g_i$ | Shorthand notation for the candidate models considered. See $g_i(x)$ |
| ML | Maximum Likelihood method of estimation (Appendix A) |
| MLE | Maximum Likelihood Estimate (or estimator) |
| $n$ | Sample size. However, some problems do not have a simple sample size as the effective sample size varies by parameter |
| QAICc | A version of AICc for overdispersed count data where quasi-likelihood adjustments are required, hence $\hat{c}$ is used |
| $\theta$ | Used to denote a generic parameter vector (such as a set of conditional survival probabilities, $S_i$ or a set of regression coefficients, $\beta_i$) |
| $\hat{\theta}$ | An estimator of the generic parameter vector $\theta$. Usually these are MLEs. The "hat" denotes an estimate or estimator, rather than the parameter value |
| $\rho_{x,y}$ | The population correlation coefficient between variables $x$ and $y$ |
| $R$ | The number of candidate hypotheses or models in the set |
| RSS | The residual sum of squares in least squares methods. Often referred to as the error sum of squares or sum of |

|  | squares due to error (SSE). The RSS is $\Sigma(e_i)^2$ for $i = 1,$ $2,\ldots, n$ |
| $\sigma^2$ | The residual variance in "regression." Here I will use the MLE of this quantity; $\sigma^2 = $ RSS $/ n$ and not the more usual "unbiased" LS estimator (i.e., RSS/$(n-K)$ |
| se or se($\hat{\theta}$) | Standard error and standard error of the estimator $\hat{\theta}$. Used as a measure of precision (or repeatability) |
| TIC | Takeuchi information criterion |
| Tr | The matrix trace operator; the sum of the diagonal elements of a square matrix |
| var($\hat{\theta}$) | The sampling variance of the estimator $\hat{\theta}$. The square root of this quantity is the standard error. Both are measures of precision |
| $w_i$ | Akaike weights. Used with any of the information criteria that are estimates of Kullback–Leibler information (e.g., AIC, AICc, QAICc, TIC). Estimates of the probability of model $i$ being the K–L best model, given the data and the model set. These are analogous to Bayesian posterior model probabilities |
| $W_+(j)$ | The sum of Akaike weights over all models that contain the explanatory variable $j$ |
| $X$ or $X$ matrix | The data or matrix of data |
| $\propto$ | A symbol meaning "proportional to" |
| $\equiv$ | A symbol meaning "defined as" |
| $\approx$ | A symbol meaning "approximately equal to" |
| $\mid$ | A symbol meaning that entities to the right are "given" or "conditional upon" or known, as in $\mathcal{L}\,(\theta\mid x)$ |
| $\ll$ | A symbol meaning "much less than" |

Definitions of other statistical terms are given by Everitt (1998).

# 1

# Introduction: Science Hypotheses and Science Philosophy

Thomas C. Chamberlin (1843–1928) was trained as a geologist but had a keen interest in and impact on science philosophy. His 1890 paper in *Science* advocated the use of "multiple working hypotheses" and is central to the information-theoretic approaches. He was the director of the Walker Museum at the University of Chicago, president of the American Association for the Advancement of Science, and the founder and editor of the *Journal of Geology*. Chamberlin was the president of the University of Wisconsin at the time the paper was prepared. The paper was republished in *Science* in 1965 and is still very worthwhile reading as much of science turned, unfortunately, to testing null hypotheses by the early part of the twentieth century.

## 1.1 Some Science Background

Science is about discovering new things, about better understanding processes and systems, and generally furthering our knowledge. Deep in science philosophy is the notion of hypotheses and mathematical models to represent these hypotheses. It is partially the quantification of hypotheses that provides the illusive concept of *rigor* in science. Science is partially an adversarial process;

hypotheses battle for primacy aided by observations, data, and models. Science is one of the few human endeavors that is truly progressive. Progress in science is defined as approaching an increased understanding of truth – science evolves in a sense.

Philosophy of science is concerned with the justification of scientific practices. For instance, it may be obvious that an experimenter would want to avoid confounding; however, it may be far less obvious why randomization or parsimony is often so critical in empirical science. Establishing causation and making proper inductive inferences are the domains of a scientist. These are deep issues where science philosophy has played an important role. A good scientist should try to further understanding of philosophy during her or his lifetime of work.

The scientific method tries to formalize, and make efficient, the everyday process of "finding things out." Good science is strategic. Science is fundamentally about understanding, not so much about decisions (however, there are many solid approaches to making "scientific decisions" but that is another subject). Aldo Leopold (1933:231) stated, "We are not trying to render a judgment, rather to qualify our minds to comprehend the meaning of evidence." We will see that evidence can be formally quantified – this is the science of the matter. However, in application, we then often want to qualify such evidence to aid comprehension. Such qualifications are value judgments, are not unique, and can be contentious.

Science is not so much about what is known (although we do speak of the "body of scientific knowledge"), as it is the process of finding out about new things. Science makes progress by providing evidence that good hypotheses are poor so that they can be replaced by even better hypotheses. Science never stops; it is always looking for more.

Ideally, perhaps scientists should be disinterested and unbiased observers. I suspect that human nature prevents this ideal in most of us; instead, we should admit that we often have some "leaning" on many subjects. This leaning reflects, partially, our interest in the subject in the first place. This predisposition can be accounted for as hypotheses are evaluated objectively, relative to one another. For example, when deriving a small set of hypotheses to be evaluated with data and models, one investigator may have a favorite hypothesis, while a colleague may favor another. This can lead to a spirit of competition for ideas and new hypotheses that can be healthy in learning. This is where good data and a sound approach to evaluating the relative strength of evidence for the set of hypotheses become fundamentally important.

Evidence is defined by the American College Dictionary as "grounds for belief" and "something that makes evident." Proof is evidence so complete and convincing as to put a conclusion beyond reasonable doubt. Strict proof may be rare in life sciences. Faith is belief without evidence.

Substantial elements of personal judgment enter in scientific research, especially in the choice of topics of study and in deeper issues of interpretation. To some extent, however, the goal of scientific methods is to minimize that personal element and subjectivity. Often, we will see that the science of a matter consists of various pieces of quantitative evidence: things like ranking of hypotheses derived *a priori*, the probability of hypothesis *j*, estimates of parameters, and a measure of their precision. Perhaps it is useful to think that science stops there. Then, value judgments can be offered to qualify the result and therefore aid in its interpretation. Such interpretations can be offered by anyone and these may be fairly similar across individuals or may vary quite substantially. The value judgments by the investigator might be of special interest; this is why a Ph.D. level of education becomes important in scientific studies. In the end, the qualification of the quantitative evidence (the science result) involves value judgment that may vary by individual. Goodman and Royall (1988:1573–1574) note:

> ...the use of evidential measures forces us to bring scientific judgment to data analysis, and shows us the difference between what the data are telling us and what we are telling ourselves.

## 1.2   Multiple Working Hypotheses

Thomas Chamberlin wrote several papers over a century ago calling for scientists to adopt what he called "multiple working hypotheses." Francis Bacon advocated a similar science strategy 400 years earlier. Their proposal is a sterling blueprint for an effective science strategy but the approach has been underused during the past century.

Under Chamberlin's strategy, one carefully derives several *plausible* science hypotheses ($H_i$) that become the entire focus of the investigation:

$$\{H_1, H_2, ..., H_R\}, \text{where } R \geq 2.$$

These hypotheses are to be well thought out and derived prior to studying the specific data and ideally prior to data collection. In his time, I believe Chamberlin was thinking that $R$ was in the 2–4 range (i.e., small). Forming a small set of plausible hypotheses is where science enters the issue and is the most important step. Research investigators need the ability to think hard about plausible explanations (hypotheses) for a system of interest. Our present science culture places too little emphasis on the derivation of multiple working hypotheses. Many scientific hypotheses seem shallow and uninteresting and most cases are a single science hypothesis to be contrasted with a "null" hypothesis. Such practice cannot be considered twenty-first century science.

Once the *a priori* set of hypotheses has been carefully defined, then one can begin to ask about their relative empirical support. Royall (1997) asks: Given the data,

> ...how do we quantify the strength of evidence for one explanation over the alternatives?

Edwards (1972) states:

*Our problem is to assess the relative merits of rival hypotheses in the light of observational or experimental data that bear upon them.*

Chamberlin (1890:758) posed the question,

*...what is the measure of probability on the one side or the other...?*

Stated another way,

*What is the empirical evidence for hypothesis j relative to the others in the set?*

These are different ways to ask the *fundamental methodological question* in empirical science. Until fairly recently, science had no general methodological approach to providing answers to these questions. Certainly, null hypothesis testing is quite distant from these serious questions. Hypotheses not in the set remain out of consideration (but more on this later). Finally, one must always consider the possibility that none of the hypotheses have any substantial merit. In such cases, more experience and thinking are required.

Chamberlin said little about models and associated quantification (modeling was the subject of contributions in the twentieth century) and he said even less about *how* the various working hypotheses might be evaluated (what we now often term "strength of evidence"). Given his education in geology, it is possible he was thinking of questions where the answer was effectively deterministic and where there was little uncertainty concerning the evidence. Effective ways are needed to provide relative *evidence* for members in this set of science hypotheses. Such ways are the focus of this text.

Chamberlin believed that the derivation of a "family of hypotheses" had special merit and by its very nature promoted thoroughness. He felt the value of working hypotheses was in its suggestiveness of lines of inquiry that might otherwise have been overlooked. This approach leads to certain "habits of mind" – special ways of thinking carefully about new problems ("thinking outside the box"). Collaboration with peers can often lead to interesting insights concerning alternatives.

## 1.3   Bovine TB Transmission in Ferrets

Caley and Hone (2002) present a nice example of hypothesis generation concerning the force of infection of bovine tuberculosis (*Mycobacterium bovis*) in feral ferrets (*Mustela furo*) in New Zealand. Caley (personal communication) provided a synthesis as to how they approached the science issue. Their analysis is quite comprehensive; I will only highlight some simple aspects to illustrate their approach to deriving multiple working hypotheses (I encourage readers to study their paper for deeper issues). Caley and Hone (2002) derived 12 alternative hypotheses concerning disease transmission; this took place over several months and they made a major effort to get an exhaustive set of hypotheses. Caley was closest to the issue and his beliefs were centered around a hypothesis of a dietary infection hazard ($H_4$ below). Hone was less close to the polarized political debate and the authors viewed this as a benefit as he facilitated a more open perspective to developing plausible alternatives.

They examined the ecological and epidemiological literature as an aid in the derivation of alternative hypotheses, but this examination was not restricted to either ferrets or bovine tuberculosis. They had both science colleagues as well as natural resource managers to debate the merits of various alternative hypotheses. Over time, the tentative hypothesis set narrowed and expanded as a result of a deliberate attempt to "think hard." The first five hypotheses (given below) became somewhat "obvious" and the seven remaining hypotheses arose from recognizing that the first five were not mutually exclusive. They eventually described about 20 hypotheses using logical combinations of these five. The framework for these hypotheses and the analysis to follow includes gender and site as factors in each case.

The most difficult issues involved decisions about the more complex hypotheses and the potential lack of uniqueness of some combinations of the five base hypotheses. Eventually, they decided on 12 hypotheses. To keep this example manageable, it will suffice to focus attention on their five base hypotheses below:

$H_1$    Transmission occurs from mother to offspring during suckling until the age of weaning, which occurs at 1.5–2.0 months of age

$H_2$    Transmission occurs during mating and fighting activities associated with it, from the age of 10 months when the breeding season starts

$H_3$    Transmission occurs during routine social activities from the age of independence, estimated to be about 2–3 months, such as sharing dens simultaneously

$H_4$    Transmission occurs during scavenging/killing tuberculosis carrion/prey from the age of weaning (1.5–2.0 months of age)

$H_5$    Transmission occurs from birth because of environmental contamination

Note that each hypothesis asks about *How* and *When*; these are often better science questions than merely *What*, as this tends to be merely descriptive. The question is not "is there an effect" rather there is interest in the size of the effect and this is measured by estimates of model parameters.

If these five hypotheses could be ranked (simple ranking is a form of evidence), based on the data, many people would realize how much more relevant this would be compared to an array of classic *P*-values. I am unsure what the null hypothesis might be: transmission is random, but that seems unlikely.

Most of us have difficulty with complex issues and some forms of quantification. Chamberlin warned that it was easier and seemingly more pleasing to think in terms of simple interpretations than to recognize and evaluate the multiple factors that may often be operating. He provided an example where he felt people like to be told that the Great Lakes basins in the United States were scooped out by glaciers, than to be taught that three or more factors working successively or simultaneously were responsible and to then try to partition the relative importance of these factors. This is an important insight, while realizing that effective theory often requires some idealization and simplification.

Scientists should think long and hard about the *a priori* hypotheses to include in the set for study and evaluation. This critical step can often take months of thinking and rethinking the issues (*a la* Caley and Hone). Oliver (1991) said it well,

> When you come across some observation that does not fit the standard explanation, let your mind wonder to see whether some radically different interpretation might do a better job. Perhaps you will think of something that will fit both the new data and the old data and thereby supplant the standard explanation. Toy with different perspectives. Look for the unusual. Try consciously to innovate. Train yourself to imagine new schemes and innovative ways to fit the pieces together. Seek the joy of discovery. Always test your new thoughts against the facts, of course, in rigorous, cold-blooded, unemotional scientific manner. But play the great game of the visionary and the innovator as well.

Ken Burnham (personal communication) advises,

> Ideally, one should have a firm justification for including certain hypotheses in the set and, conversely, have an equally firm justification for excluding other hypotheses from the set.

The definition of the hypotheses in the set is perhaps the most important part of the investigation. This set defines the science at the moment. Statistical science is most successful when full attention is given to problem formulation and hypothesizing creative, plausible alternatives. This is often the case where "two heads are better than one" and where there is a competition for ideas and alternatives. Ball et al. (2005) provide a nice example where 15 hypotheses were developed *a priori* concerning predictions about vegetation and substrate affinities for Palm Springs ground squirrel (*Spermophilus tereticaudus chlorus*).

## 1.4    Approaches to Scientific Investigations

Much of both science and statistics is about *inductive inferences*. This is a formal process whereby a conclusion about a sample is extrapolated to the population from which the sample was drawn. The data come from the sample only; the remaining members of the population are not observed.

Inductive inference can also be thought of as a conclusion from the past about the future as in forecasting or prediction. Inference is an act or a process.

For such inferences to be valid, in principle, assumptions must be met; e.g., some type of probabilistic sampling of the well-defined population. Furthermore, there are results from logic stating that there is always uncertainty in making inductive inferences. This uncertainty leads to the need to carefully quantify the uncertainty of such inferences (e.g., variances, covariances, standard errors, various types of confidence intervals) and worry about possible biases.

Inference in many of the life sciences is challenging because of the inherent *variation* in living systems. In addition, there are often multiple causal factors and, thus, the need for replication, controls, and worry about confounding.

## 1.4.1   Experimental Studies

Experiments are the Holy Grail of science because they allow inferences about causation. In science, the word experiment implies treatment vs. control groups, where experimental units are randomly assigned to these groups, and there is deliberate replication. Anything less than these three conditions should not properly be called an experiment. In cases where random assignment has not been done (but treatment and control groups are defined and replication is in place), they are often called "quasiexperiments" and there is a large literature on this important case. Studies without a control group are likely to yield disappointing results as the effect size cannot be estimated (there are exceptions), while studies without replication yield results that are usually tenuous at best.

**The main purpose of an experiment is to estimate the size of the effect on a response variable of interest *caused* by the treatment**. This is primarily an estimation problem: One wants to have an estimate of the effect size and its standard error (or some other appropriate measure of precision). The experimenter wants to know something about the effect of the treatment on some response variable: Is the effect trivial, small, medium, large, or extra large? This has little to do with *testing* the null hypothesis that the treatment had exactly zero effect. In these cases, I find relatively little use for extensive tables of sums of squares, mean squares, F statistics, various degrees of freedom, and the ever-present *P*-values, followed by a decision to "reject" or "fail to reject" the null hypothesis.

Even in strict experiments, the null is almost never particularly plausible; it is the size of the effect caused by the treatment that is of scientific interest. For example, "I fed bison calves in the treatment group a special dietary supplement and, at that dose level, it caused them to gain an average of six pounds per week over that of the calves in the control group fed the same volume or weight." This science finding might be followed by an evaluation of costs of the supplement and other factors thought to be relevant. If ordered treatments are part of the experimental design, then the "effect size" becomes the nature of the (causal) functional response.

A great deal of excellent information exists concerning the design and conduct of experiments and the analysis of experimental data. It is often useful to think of experiments as "design based inference" as the *design* stipulates a (single) model and the causal inference stems from the experimental design. For example, a randomized complete block design implies (only) a two-way ANOVA model and subsequent analysis. This is an area of statistical science that is quite mature – hundreds of books extol its virtues; it is usually well taught, and a huge variety of computer software exists for this. If the scientific situation allows, experiments are highly regarded and recommended – they represent a philosophical "gold standard" of scientific endeavor because they address causation, not merely association or correlation.

There are close links between experimental design and sampling design (see Snedecor and Cochran 1989). While the objectives clearly differ, the estimators can be viewed within a common statistical framework.

## 1.4.2   Descriptive Studies

The harsh reality, in many cases, is that a strict experiment simply cannot be done. A quick survey of journals such as *Ecology, Journal of Animal Ecology*, or *Journal of Conservation Biology* will reveal many papers that are not about experimental results. In some cases, an experiment could have been done but the investigators did not realize this and then the results are nearly always compromised. In most cases, however, there are a host of valid reasons why a strict experiment is not feasible. Ethical concerns often prevent strict experimentation in human medical research. In many cases, investigators turn to descriptive work. Descriptive work certainly has its place in science but such inferences are (or should be thought of as) more shallow and tentative.

Related to what I call descriptive studies is the notion of "exploratory data analysis" and much has been written about this approach. I believe too much emphasis has been placed on descriptive work. I also believe that some types of exploratory data analysis are a relatively poor way to make rapid progress in the empirical sciences. It is too easy to mistakenly think that results from *post hoc* analyses (data dredging) will lead to interesting new hypotheses, when often such results have high probabilities of being, in fact, spurious. The best ways to obtain interesting alternative hypotheses is to think, read, study, attend scientific meetings, and communicate with both colleagues and rivals.

## 1.4.3   Confirmatory Studies

Another alternative, confirmatory investigation, lies in between strict experiments that may provide evidence of causation and the descriptive studies that often provide only "what?" Confirmatory investigations begin by hypothesizing alternatives prior to data analysis and, ideally, even before data collection. When data are analyzed and "results" appear, these are confirming prior hypotheses. This is a level above descriptive studies where findings come almost by surprise in many cases. The investigator thinks, "Wow, who would have thought that nest success of species X was related to estimated cloud cover on the day and time an observer found the nest." On the contrary, under a confirmatory format, the investigator notes something like, "Oh, we have thought for some time that nest success is influenced by concealment at the time of nest initiation, now we have some (confirmatory) evidence of this." Better still, "now we ask some hypothetical questions about why and when concealment is important."

Clearly, there is a close link between the confirmatory approach and Chamberlin's ideas of multiple working hypotheses. Platt's (1964) well-known paper on "strong inference" is closely related and addresses the issue of strategy in science. One still lacks a notion of strict causation but the strength of evidence is nearly always above that for descriptive studies. The only price to be paid to achieve a confirmatory result is *a priori* thinking. We must ask why more confirmatory research is not being done. A little hard thinking before data collection and analysis provides a much stronger inference. This approach sets up a basis for formal evidence. Putting in place the *a priori* hypotheses is just good science procedure and it yields a superior result. Of course, some *post hoc* analysis can be done, and I promote this, but these inferences must be treated with appropriate caution. The reader should be informed which results were from the confirmatory process and which came from *post hoc* analyses of the same data. Confirmatory investigations are the domain of model based inference and the primary focus of this text.

Fundamental differences between strict experiments and other types of studies can be further understood in terms of residual variation (e.g., the $\epsilon_i$ in regression). In experimental data, the residuals are the component of variation that is considered to be random, where the model is defined by the experimental design, and, in this sense, is "known." In contrast, nonexperimental (observational) data must also (i.e., in addition to) treat the residuals as containing the effect of as yet unknown confounding variables on the estimated response variable, (somehow) given the model. Here, the model is not known and must be estimated using some model selection procedure. Strict experimentation is quite distinct from other approaches to science questions.

I find that many investigators have a fear during data analysis that they will miss an effect that is real and perhaps important. They worry that the data are wanting to tell something, but that they will miss this finding by "not looking hard enough." This is a valid fear and it may be common, in complex settings, that some second-order effects are missed, even though they might be in the data. Perhaps an interaction is missed or a nonlinearity is left unnoticed. Investigators try to minimize missing effects by examining "all possible models" or using some multivariate software – there are huge inferential risks associated with these seemingly logical approaches.

The associated risk is finding a spurious effect. That is, the analysis picks up some effect that is particular to the data set that is not part of the process of interest. In a sense, noise is being detected and modeled as if it were part of the process. One has no way to know, based on a given data set of limited size and scope, if a particular effect is spurious or real. This is the risk that people tend to forget and misunderstand. They want to be sure they "did not miss anything" and while doing so they, instead, find results that are spurious. Many things can be done to lessen this nasty issue, but too many investigators continue to forge ahead, unaware of the risks involved. Spurious results arise with high probability when one has little subject matter theory, measured many variables (e.g., more than the sample size in extreme cases), had small sample size (e.g., 20 or 50), and many models (hundreds, thousands, or even millions of models). It takes some experience and maturity to really begin to understand these issues. Model selection theory can help pinpoint these problems but by then it may be too late to salvage the study.

## 1.5 Science Hypothesis Set Evolves

Expanding on the ideas of Chamberlin and Platt, we want the set of hypotheses to evolve over time. That is, a team of researchers might start with a set of five hypotheses and find, after a careful empirical evaluation, that two of these were implausible to the point they could be dropped from further consideration. Thus, the set is reduced to only three hypotheses that survived the evaluation. These three might then be further refined and elaborated upon and perhaps one new hypothesis introduced. Hence, a set of four hypotheses are now available for evaluation with new data. If the data set is fairly small, one must be careful not to discard more complex hypotheses, particularly if a much superior data set is expected for the next evaluation. Science can progress very rapidly as the evolving set of hypotheses are constantly challenged with new data (information) and careful evaluation.

The hypothesis set is made to evolve over months, years, and decades; the goal is to keep careful focus on the hypotheses that remain plausible and in the set. It is these hypotheses where improved understanding is sought. The strategy is to constantly make this set move along as knowledge is broadened and further understanding is gained. *A priori* reasoning and hard thinking are both critical and difficult. Scientists should not fail to acknowledge that there may be more than one process that would yield a particular outcome (Platt 1964; Pigliucci 2002a). A real focus needs to be placed on the addition of very new and different hypotheses as the next set is defined. Ideally, the model set would be built to consider those outcomes using experiments or observational studies to separate the alternative hypotheses.

The hypothesis set might ideally evolve as national or international teams vie for understanding and knowledge. Peer pressure and national pride might help drive progress on some interesting problems by showing that one or more hypotheses are implausible, relative to the others in the set. By then other teams are already formulating new hypotheses, perhaps suggested by the prior results. Pressure to evolve the set might be within a laboratory, university, or an agency, where fast learning is important. Of course, one's own personal scientific progress can be based on the notion of wanting the set to evolve so that rapid understanding may be achieved.

One must not be too eager to rule a particular hypothesis "implausible"; if there are seemingly valid reasons to retain it, then its retention may be appropriate. The decision to retain it must be based on the quantitative evidence and ways to obtain such evidence are given in the following chapters. Sometimes a complex hypothesis is rejected largely because the data set is small and there is little information in the data. In this case, the hypothesis should probably be retained if the next data set is thought to be larger and more informative.

## 1.6   Null Hypothesis Testing

Soon after Chamberlin's science strategy was published in 1890 and widely accepted, investigators found it easy to derive a single science hypothesis ($H_a$) and then contrast it with a "null" hypothesis ($H_0$),

$$\text{Null hypothesis } H_0 \text{ vs. Science hypothesis } H_a.$$

This approach was prompted by emerging developments for randomization and strict experiments in statistics in the early 1900s. "Student," Fisher, Neyman, Pearson, Wald, and many other pioneers in statistical theory developed methods for "testing" null hypotheses and this became the dominant analysis paradigm for perhaps seven to nine decades. The approach advocated by Fisher differed substantially from that of Neyman and Pearson and this led to a heated and protracted debate. Now, these approaches are combined in a fashion that would have greatly displeased the combatants involved. Such testing has come under increasingly harsh criticism since at least 1938, particularly when used for the analysis of data from observational studies. It now seems clear that this standard "testing" approach is of limited value as new approaches have several important advantages. Still, I find statistics departments around the world teaching primarily null hypothesis testing methods developed in the early 1900s. Worse yet perhaps is the continued focus on the myriad approaches to multiple comparison tests. This focus on relatively poor methods seems increasingly senseless and few people seem to have any idea why good statistics departments cannot do better in teaching current theory and application (e.g., likelihood, information-theoretic, and objective Bayesian approaches).

It is important to realize that null hypothesis testing was *not* what Chamberlin wanted or advocated. We so often conclude, essentially, "We rejected the null hypothesis that was uninteresting or implausible in the first place, $P < 0.05$." Chamberlin wanted an *array* of *plausible* hypotheses derived and subjected to careful evaluation. We often fail to fault the trivial null hypotheses so often published in scientific journals. In most cases, the null hypothesis is hardly plausible and this makes the study vacuous from the outset. Chamberlin noted, "The vitality of the study quickly disappears when the object sought is a mere collection of dead, unmeaning facts." For example,

$H_0$:    the population size of species $X$ is the same in urban and rural areas,
$H_0$:    species diversity does not change through geologic time,
$H_0$:    the population correlation between variables $X$ and $Y$ is exactly 0, and
$H_0$:    bears do not go in the woods.

Surely, these null hypotheses are false on simple *a priori* grounds – data collection and analysis are hardly needed in cases such as these. A hundred thousand such null hypothesis tests have appeared in the journal *Ecology* in the recent past (over a 20-year period, Anderson et al. 2000). C. R. Rao (2004), the famous Indian statistician, recently said it well, "... in current practice of testing a null hypothesis, we are asking the wrong question and getting a confusing answer."

We must encourage and reward hard thinking. There must be a premium placed on thinking, innovation, synthesis, and creativity; the computer will be the last to know! We need to ask more about *how*, *when*, and *why*, which are more interesting and potentially important, instead of such a focus on *what*, which is often only descriptive.

## 1.7   Evidence and Inferences

Scientific evidence lies in a triangular plane surrounded by philosophy, statistics, and subject matter science. Statistical inference is the foundation of modern scientific thinking. Many people are unaware of the extent that fundamental scientific thought processes have been influenced by philosophers. This may be especially true in life sciences.

Evidence is something less than proof! Evidence provides a foundation leading to understanding and this ideally leads to useful theory. Theory is a body of knowledge and understanding that has stood the test of considerable time and effort to disprove it. A theory makes predictions and has been found to be generally useful. Theory might eventually be accepted as a law.

Chamberlin (1890) asked, "What is the measure of probability on one side or the other." Methodology to allow such probabilities took nearly 100 years to develop! The field of statistics became sidetracked and much of the twentieth century was occupied with "testing" null hypotheses and resulting $P$-values that were often used as if they represented a strength of evidence. I believe that $P$-values reject or fail to reject dichotomies, and the often trivial null hypotheses that they represent are being replaced with formal methods to quantify and allow comprehension of the evidence for members of a set of alternative science hypotheses.

Examination of the differing probabilities shows a stark difference in meaning. A traditional null hypothesis test is based on a summarization of the data into an appropriate test statistic $T$. The associated $P$-value is the probability of $T$ being as large or larger, *given* the null hypothesis is true. The $P$-value is a so-called "tail probability" and has been criticized as assigning probability to data not collected. Shortening the definition slightly, the $P$-value is

**Prob(observed data or more extreme | null hypothesis).**

Conditioning on the null hypothesis might seem odd as it is uncommon to believe in the null. If most null hypotheses seem implausible on *a priori* grounds, why condition on such a notion? I think one should condition on the data – that is why data are collected. Data serve as the arbitrator, the jury, the judge; conditioning on the null hypothesis is not intuitive. The relevant probabilities deal directly with the individual science hypotheses, *given* the data,

$$\textbf{Prob}(H_j \,|\, \textbf{data}), \quad \text{for } j = 1, 2, \ldots, R.$$

## 1.8   Hardening of Portland Cement

I will use a well-known problem involving cement hardening as an example. This is a relatively simple problem that can be made to illustrate several important issues and I will use it in several chapters to follow. Woods et al. (1932:635–649) published the results of a small study of the hardening of Portland cement; Daniel and Wood (1971) and Burnham and Anderson (2002) provide further details on these data for the interested reader. Interest was in the calories of heat evolved per gram of cement after 180 days; this relates to hardening and was their response variable, denoted here as $y$. The objective of the study was twofold: (1) identify the important variables related to the response variable and (2) use these to predict the response variable. Four predictor (or explanatory) variables were of interest:

$x_1 = \%$ calcium aluminate ($3CaO \cdot Al_2O_3$)
$x_2 = \%$ tricalcium silicate ($3CaO \cdot SiO_2$)
$x_3 = \%$ tetracalcium alumino ferrite ($4CaO \cdot Al_2O_3 \cdot Fe_2O_3$)
$x_4 = \%$ dicalcium silicate ($2CaO \cdot SiO_2$)

In arriving at a set of plausible hypotheses, several lines of reasoning could be followed. The first observation might be that cement is a mixture of ingredients, thus hypotheses involving only a single variable might be deemed implausible. A more shrewd observation might be that $x_2$ and $x_4$ have very similar chemical make up,

$$x_2 \quad \text{tricalcium silicate} \quad (3\text{CaO} \cdot \text{SiO}_2),$$
$$x_4 \quad \text{dicalcium silicate} \quad (2\text{CaO} \cdot \text{SiO}_2).$$

Also, the chemical composition of variables $x_1$ and $x_3$ also seem somewhat similar,

$$x_1 \quad \text{calcium aluminate} \qquad\qquad (3\text{CaO} \cdot \text{Al}_2\text{O}_3),$$
$$x_3 \quad \text{tetracalcium alumino ferrite} \quad (4\text{CaO} \cdot \text{Al}_2\text{O}_3 \cdot \text{Fe}_2\text{O}_3).$$

This line of thinking might avoid hypotheses involving the pair of variables $x_2$ and $x_4$ and the pair $x_1$ and $x_3$ as they are so similar. We can further check these issues when we have data available (i.e., look at the sampling correlations between these pairs of variables). The investigator might consider the importance of an interaction such as $x_1 * x_2$ and add these as hypotheses. A skeptic (scientists should be skeptics) might ponder the notion that none of the four predictor variables had any merit; if this seems plausible, then it should be included in the hypothesis set. This would be the intercept only model (a null model of sorts) with two parameters: the mean $\beta_0$ and the residual variance $\sigma^2$. Here this seems unlikely if we assume people in the 1930s had some basic notion of what they were doing in regard to cement making. Still, if this is deemed plausible, it should go in the set; if not, then it should be excluded. We will leave it just for this example (ordinarily I would deem this implausible in this case and omit it). In summary, we might have the following five hypotheses in the set (defined *a priori*; i.e., before data analysis):

$H_1$   No variables
$H_2$   $x_1$ and $x_2$
$H_3$   $x_1$ and $x_2$ and $x_1 * x_2$
$H_4$   $x_3$ and $x_4$
$H_5$   $x_3$ and $x_4$ and $x_3 * x_4$

Data and simple models of these hypotheses will be given in Chap. 2.

## 1.9   What Does Science Try to Provide?

There are several common goals in scientific inquiry. Science is very broad and actually defies a simple but adequate definition. Science is a process leading to discovery, understanding, and solutions of well-defined questions about effects; some strategies are better than others. Some parts of science might be classified as:

- Evaluating the strength of evidence for alternative science hypotheses, *a la* Chamberlin
- Prediction of an outcome, given data, a model, estimates of model parameters, and specific values of the predictor variables

- Determination of model structure (e.g., concave or convex in a simple setting)
- Selection of "important" variables from a larger set (variable selection in regression or discriminate function analysis)
- Pattern recognition or smoothing (parsimony)

All of these classes are examples of model based inference. I will address these matters in the material to follow.

There are some distinctions between science and technology that are sometimes worth noting. Estimates of effect size or predictions might often be best classed as technology. Science might best be thought of as discovery in an exciting sense; for example, providing a strength of evidence for fundamental alternatives. Classification of science vs. technology is rarely distinct and there are wide areas of overlap; still the distinction is often useful to keep in mind.

## 1.10   Remarks

Developing interesting hypotheses is an art but people can become adept at this with dedication and practice. Ford (2000:Chaps. 4 and 13) and Gotelli and Ellison (2004:Chap. 4) provide relevant reading. Cox (1990, 1995) reviews the relationships between hypothesizing and modeling. Krebs (2000) offers a relevant and easy-to-read review of hypotheses and models. Peirce (1955), Moore and Parker (1986), Abelson (1995), Pigliucci (2002b), and Cohen and Medley (2005) give valuable perspectives on hard thinking, statistical reasoning, and science principles. Careful reading of journal papers in one's field can be enlightening as you can begin to understand how others thought about their science. However, only a minority of papers are exemplary in this important regard. Beyond reading one must think broadly about alternatives, draw from conversations with colleagues, attend conferences, and use new technologies (e.g., the Internet and e-mail) to forge science ideas.

Mead (1988) and Manly (1992) provide methods for the design and analysis of experimental data. Cook and Campbell (1979) and Shadish et al. (2002) deal with quasiexperiments, both design and analysis, in readable books. Observational studies are well covered by Rosenbaum (2002). There are many dozens of good books on experimentation and applied statistics. I will let the readers make their own choices; however, I find that Resetarites and Bernardo (1998) and Williams et al. (2002) make many deeper issues clear with examples.

It might seem surprising but during Chamberlin's time there was an effort to minimize theorizing; this dead end was an attempt to reform "ruling theory" that was in place then. Chamberlin believed his strategy "promotes thoroughness and suggests line of inquiry that might otherwise be overlooked." Additional insights on Chamberlin's method are found in Elliott and Brook (2007).

Many science problems arise where the objective is a structural issue (see Blanckenhorn et al. 2004 for an evolutionary issue that focuses on linearity vs. nonlinearity).

Rao's (2004) short comment is full of interesting insights and Krebs (2000) provides a readable treatment concerning hypothesizing. Forsche's (1963) one-page paper in *Science* is delightful reading. More troublesome is O'Connor's (2000) paper on lack of progress in ecology compared to other areas of biology (also see the interesting follow-up by Swihart et al. 2002). However, see Mauer (1999) for evidence of substantial progress.

Kendall and Gould (2002) respond to the criticism that statistics departments often provide poor course material for people in the life sciences. Their excellent point is that biologists often arrive for statistics courses with a poor science background! Biology students arrive without a grounding in experimentation, causation, confounding, and other subjects, making statistical concepts seem out of context (because students often lack that context). Biology students should have a better grounding in the history of science in general and in their field in particular. Science philosophy is equally fundamental in university courses.

Anderson et al. (2001a) provide an overview of the issue of spurious effects and how to minimize this risk. Hobbs and Hilborn (2006) examine alternatives to null hypothesis testing in a readable paper aimed at ecologists. Their Fig. 1 is interesting to compare with the results shown by Anderson et al. (2000).

O'Connor (2000) states: "Moreover, many of the critical breakthroughs in molecular biology have come from experiments that discriminate between alternative outcomes. Critically, ecology seems to have substituted the statement of what will be observed as the hypothesis."

Freedman (1983) used MC simulation methods and stepwise regression to illustrate the difficulties faced when one has (1) little or no theory, (2) a large number of "models" (as there is little theory leading to *a priori* hypotheses, and (3) small sample size. In such cases, inference is very risky and a plethora of spurious results are found (also see Flack and Chang 1987, Freedman 1983, and Rencher and Pun 1980). Freedman demonstrated this phenomenon by a large matrix of uncorrelated random numbers and stepwise regression. While there were clearly no relationships underlying the data, numerous "significant findings" resulted. This has become known as Freedman's paradox. Unfortunately, many studies in the life sciences are done under these conditions and find their way into the published literature.

The paper by Goodman and Royall (1988) contains many philosophical insights and are well worth careful reading. Recently, Keppie (2006:244) provides fresh perspectives, including mention of the "...temptation to advocate value beyond evidence." Hilborn and Mangle (1997) and Kuhn (1970) contain valuable insights.

There are many good books on science philosophy; I enjoy Horner and Westacott (2000), and Ford's (2000) book is a standard one. Papers by Platt (1964) and Popper (1972) are highly recommended. Taper and Lele (2004) summarize several philosophies of interest. Platt (1964) is quoted by Pigliucci

(2002:92) saying, "Scientists become method oriented rather than problem oriented. Stop doing experiments for a while and think."

## 1.11 Exercises

The following exercises are provided to strengthen the understanding of some of the conceptual issues in this introductory chapter. These questions can be addressed individually but I find it more effective and fun to tackle these issues in small groups of people. This assumes everyone in the group has read the paper and is ready to think hard about the issues.

1. Obtain a good scientific journal in your subdiscipline of interest and look for an article where the investigators used a confirmatory approach or model based inference and begin to carefully critique it. Consider the following questions:

   a. Were only two alternatives hypothesized? Or, were there more alternatives hypothesized?
   b. In any case, do you consider these to be plausible?
   c. Did the investigators justify their set of alternative science hypotheses or was there a rush to models?
   d. Was there a clear statement of the *a priori* hypotheses in the set?
   e. Did the authors entertain all possible models? What problems might this cause (advanced question)?
   f. Did the authors clearly present the body of evidence to the reader?
   g. Did they then offer some qualification (value judgment) of the evidence to aid in interpretation?
   h. What alternative hypotheses would you have added? Before data collection and analysis? After data analysis (*post hoc*)?
   i. As an associate editor, what would be your basis for rejection of this paper (had this been a submitted manuscript)?
   j. As a reviewer, what constructive advice would you have given the authors of the submitted manuscript?
   k. What other issues might be considered in your critique?

2. Using the same scientific journal as above, find a paper in your subdiscipline of interest that used null hypothesis testing as the basis for the results. Consider the following questions:

   a. Isolate the null hypotheses tested (these are often not stated explicitly). Can you offer your judgment as to the plausibility of these null hypotheses? That is, before data analysis or reading the *Results* section of the paper, were the null hypotheses to be tested plausible? Interesting?
   b. Were estimates of effect size given with a measure of their precision? If not, can you explain this omission?

    c. Define exactly what is meant by $P$-value. Did the author use this definition correctly or (possibly inadvertently) redefine it in an *ad hoc* manner? How are data involved in $P$-value? Is $P$-value an estimate?

    d. Which hypothesis is being formally "tested"? Is it $H_0$ or $H_a$? Why?

    e. Is $P$-value a measure of strength of evidence?

    f. Do the authors provide qualitative statements concerning "significance"? Do they differentiate statistical vs. biological significance?

    g. Was a causal result implied or claimed? Do you feel this was justified? How? Why?

    h. In what sense were the units (e.g., experimental, observational, sample) taken from a well-defined sample from a population studied? This issue relates to the proper scope of inference. A very large number of similar questions could be asked here. What other issues might you consider?

3. Using a journal of choice in your field of interest, find a paper that provides the results of an experiment. Read it carefully and consider the following questions:

    a. Were there clearly defined treatment and control groups?

    b. How much replication was used? Was the sample size given?

    c. Were the experimental units randomly assigned to treatment vs. control groups? How, specifically?

    d. Did the authors imply a causal result? Does this seem justified to you?

    e. Did the authors provide an estimate of the size of the effect caused by the treatment? And some measure of its precision?

    f. How could the experiment been better (list 3–4 ways)?

    g. Assuming that the answer to at least one of the questions a, b, or c was negative, what are the consequences? Elaborate. What was lost? What could have been done differently? Is the result still of interest, in your opinion?

4. Cohen (1966, 1967, 1968) used a combination of imagination and modeling leading to several science hypotheses about optimizing reproduction of a desert plant species in randomly varying environments. This set of papers triggered perhaps dozens of field experiments to gain further insights into this issue. This set of papers would make a great brown bag discussion for those interested in developing sophisticated alternative hypotheses.

# 2

# Data and Models

Ludwig Eduard Boltzmann (1844–1906) was one of the most famous scientists of his time and he made incredible contributions in theoretical physics. He received his doctorate in 1866; most of his work was done in Austria, but he spent some years in Germany. He became full professor of mathematical physics at the University of Graz, Austria, at the age of 25. His mathematical expression for entropy was of fundamental importance throughout many areas of science. The negative of Boltzmann's entropy is a measure of "information" derived over half a century later by Kullback and Leibler. J. Bronowski wrote that Boltzmann was "an irascible, extraordinary man, an early follower of Darwin, quarrelsome and delightful, and everything that a human should be." Several books chronicle the life of this great science figure, including Cohen and Thirring (1973) and Broda (1983) and his collected technical papers appear in Hasenöhrl (1909).

## 2.1  Data

Data should be taken from an appropriate probabilistic sampling protocol or from a valid experimental design, which also involves a probabilistic component. These are important steps leading to a degree of scientific rigor. Such data often arise from probabilistic sampling of some kind and are said to be "representative." Outside of this desirable framework lie populations where such ideal sampling is largely unfeasible. For example, human populations are often composed of members that are heterogeneous to sampling. Thus, by definition, it is impossible to draw a random sample and such heterogeneity can lead to negative biases in estimators of population size. Estimators that are robust to such heterogeneity have been developed and these approaches have proven to be useful, but the standard error is often large. In general, care must be exercised to either achieve reasonably representative samples or derive models and estimators that can provide useful inferences from (the sometimes unavoidable) nonrandom sampling.

Unfortunately, it has been common in some subdisciplines to take data via what has been called "convenience sampling," that is, data are taken from roads or sidewalks or in other "convenient" ways (e.g., near a parking lot or under the shade of a tree). I believe these approaches violate accepted science practice; certainly there is not a valid basis for an inductive inference. All that might be validly said is something about only the sample itself. For example, a conclusion might be "I counted the number of birds I saw along 12 roads in western Ohio and 10% were raptors. Here, nothing can be said about birds in western Ohio in general or about the percent of the birds that were raptors as an inductive inference to some well-defined population. This situation is little different than a child who reports, "I saw some squirrels" – this sort of activity never seems to lead to a new theory or an important discovery. We know a great deal about proper data collection. There are dozens of books on sampling protocols and experimental designs. There is little excuse for getting this issue seriously wrong.

Another common error is the use of the so-called index values as the response variable. Under this approach, the response variable of interest is not recorded, rather it is replaced by a crude index value. Such index values are usually a raw count or some sort of averaging of such counts. These numbers are recorded and "analyzed." Much has been written about the use of index values and I think the evidence is conclusive that they represent an amateur, unthinking approach and is not scientific. The word "data" has the connotation that there is recoverable information in the data; index values are not data, they are just numbers. None of the procedures in this primer claim to make sense out of nonsense. If the data have not been taken with care, using proper procedures, then the so-called findings will likely be only an assortment of uncertainty and disinformation. DeLury (1947), a famous fisheries biologist, asked, "Is an untrustworthy estimate better than none?" Meaningful data of sufficient quantity are the grist of scientific bread.

There are two conceptual aspects. First, is the study sound so that an inductive inference can be justified? Second, are the data analysis methods sound? The first is not a data analysis issue, rather this question asks if the science of the matter is reasonably well in place and if the data have been collected in a reasonable manner. The second relies on adequate modeling and on objective approaches to model selection (Chap. 3). We must try to guard against rushing too quickly to data analysis, when the subject matter science is still underdeveloped or if the data are seriously compromised. The science question should be carefully thought out and plausible hypotheses derived. These matters represent hard work and must typically take thought over a period of many weeks or months. The success, in the end, will rest on these science issues being well done – we must not think the analysis will somehow make up for serious inadequacies during these initial steps. These issues will never live up to the ideal; thus, the concept of evolving sets of hypotheses often prove very useful and lead to an effective strategy for fast learning.

In serious work, data are carefully collected during a pilot study. The pilot study allows investigators the chance to work out the bugs in field or laboratory application and attempt some degree of optimization of the sampling protocol or experimental design to be used. Required sample sizes are estimated, stratification is considered, etc. Engineers routinely conduct feasibility studies before they begin a project and life scientists should take a similar approach before the actual data collection begins. If resources are found to be inadequate for the task, it is often better to wait until the needed resources are assembled before beginning the project. Such waiting allows time for additional planning and refinement while gathering the resources needed.

If data are collected in an appropriate manner, then there is *information* in the sample data about the process or system under study. In simple cases with continuous data, some of this information can be retrieved and understood using graphs (e.g., $X$ vs. $Y$), plots, histograms, or elementary descriptive statistics (e.g., estimated means and standard deviations). However, in nearly all interesting cases, a mathematical model is required to retrieve the information in the data and allow some understanding of the system. Scientists often want to make a formal inductive inference; that is, the process of going from the sample data to an inference about the population from which the sample was drawn. Deductive statements can usually be classed as either valid or invalid. However, inductive statements (inferences) are not made with certainty and inferential statements can range from very weak to very strong. Inductive inference concerns weighing evidence and judging likelihood, not proof itself. Statistical science has allowed the inductive process more rigor and the ability to address a deeper level of complexity. Nearly all questions in life sciences seem to be inductive.

Valid inductive inference is another example of rigor in science but it rests on certain important requirements. This leads to "model based inference" and this has had a long history in the sciences. The *inference* comes from a model that approximates the system or process of interest. In some sense we can think of a properly selected model as the inference (at least for inferences made from a single model).

Investigators should continue to think and rethink the collection of working hypotheses as the data are collected. This is a place to delete a hypothesis in the set because of field evidence or because it seems unfeasible to measure variables that are central to a particular hypothesis. This is a time to add new hypotheses or refine existing hypotheses. At any given time, our knowledge is based on hypotheses that have shown their competitive fitness by surviving to this point; a "survival of the fittest" as hypotheses struggle for continued existence. There is a competitive struggle that eliminates hypotheses that are unfit (found to be implausible, based on one or more data sets). For example, perhaps elevation is not important to a response variable, as first thought, rather it is actually temperature (which is negatively correlated with elevation). Should some interaction terms be considered due to observations in the laboratory? Does it seem that two variables might be very negatively correlated, suggesting care will be needed to understand this? The focus should remain on the candidate set of science hypotheses; ideally, these should be fixed once the analysis begins. Following these activities, some tentative hypotheses might be added *post hoc*, but such results must be treated more carefully.

Some common sense and art are involved in the concept of an evolving set of hypotheses. Sometimes it might be premature to delete some hypotheses if the data set is small; in such cases perhaps judgment should be reserved until another data set is available. In analyzing data from small samples, one must guard against dismissing some larger models with more structure because a new and larger data set might be able to support the additional structure. Such judgments can be guided by methods given in Chap. 4.

## 2.1.1   Hardening of Portland Cement Data

Our first example will be the data on four explanatory variables thought to be related to cement hardening. The meager ($n = 13$) data are shown in Table 2.1.

TABLE 2.1.  Cement hardening data from Woods et al. (1932). Four variables (in percent), $x_1$ = calcium aluminate ($3CaO \cdot Al_2O_3$), $x_2$ = tricalcium silicate ($3CaO \cdot SiO_2$), $x_3$ = tetracalcium alumino ferrite ($4CaO \cdot Al_2O_3 \cdot Fe_2O_3$), $x_4$ = dicalcium silicate ($2CaO \cdot SiO_2$), are given with the response variable, $y$ = calories of heat evolved per gram of cement after 180 days of hardening.

| $x_1$ | $x_2$ | $x_3$ | $x_4$ | Y |
|---|---|---|---|---|
| 7 | 26 | 6 | 60 | 78.5 |
| 1 | 29 | 15 | 52 | 74.3 |
| 11 | 56 | 8 | 20 | 104.3 |
| 11 | 31 | 8 | 47 | 87.6 |
| 7 | 52 | 6 | 33 | 95.9 |
| 11 | 55 | 9 | 22 | 109.2 |
| 3 | 71 | 17 | 6 | 102.7 |
| 1 | 31 | 22 | 44 | 72.5 |
| 2 | 54 | 18 | 22 | 93.1 |
| 21 | 47 | 4 | 26 | 115.9 |
| 1 | 40 | 23 | 34 | 83.8 |
| 11 | 66 | 9 | 12 | 113.3 |
| 10 | 68 | 8 | 12 | 109.4 |

In this case, the sample size is 13 and this must be considered to be generally inadequate. One is taught in STAT101 that a sample of 25–30 is often required just to estimate a simple mean from a sample of a population that is approximately normally distributed. Kutner et al. (2004) recommend 7–10 observations for each predictor variable. Thus, we have to be realistic when our interest lies in the estimation of more complicated parameters such as finite rates of population change ($\lambda_i$), or an enzyme inhibition rate ($f$), or some hazard rate ($h_{(t)}$). The cement hardening process is likely to be somewhat deterministic with a fairly weak stochastic component. Thus, even with the small sample available, perhaps some interesting insights can be found in this example. Note that both the response variable and the predictor variables are continuous; the response variable is unbounded while the predictor variables are percentages and bounded between 0 and 100.

The basis for a valid inductive inference in this example rests on the various chemical compounds being reasonably uniform. That is, dicalcium silicate ($2CaO \cdot SiO_2$) is the "same" from place to place. Thus, random samples would produce little variation in this variable or, for that matter, the other three variables. This is an important step or generalized inference from these data will be compromised.

Large sample size conveys many important, but sometimes subtle, advantages in the statistical sciences. Large sample size carries more information and such information is a major focus of this primer. Investigators should make every attempt to garner the resources to allow an adequate sample size to be realized. There is a large literature on the establishment of sample size, given either some background data from a small pilot survey or outright considered guesses about the system to be studied (see Eng 2004). Monte Carlo simulation studies provide another means to predict the sample size for a particular application (see Muthen and Muthen 2002).

### 2.1.2 Bovine TB Transmission in Ferrets

The second example is the data on disease transmission in ferrets in New Zealand (Table 2.2). While the sample size here is moderate ($n = 319$), estimates of the per year force of infection ($\hat{\lambda}$) varies by a factor of 88; thus it seems realistic that the data might be adequate to reveal some interesting insights. High variances seem to be the rule in many areas of life sciences, making data analysis challenging and making inferences somewhat tentative in many cases because of the uncertainty. Increased sample size can often help combat these issues.

These data consist of counts and are therefore of a substantially different type than the data in the earlier example. Ferrets were caught in baited traps systematically placed in selected areas (based on prior information from wildlife surveys or from tuberculin testing of cattle herds). Traps were checked daily over a 5–10-day sampling period. We might ask if the data came from a strict probabilistic sampling frame – no, probably not. Animals willingly

TABLE 2.2.   The data on infection and the estimated force of infection ($\hat{\lambda}$) of *Mycobacterium bovis* infection using modified exponential models (from Caley and Hone (2002).

| Site | Gender | No. examined | No. infected | $\hat{\lambda}$/ year |
|------|--------|--------------|--------------|------------------------|
| Lake Ohau | M | 57 | 3 | 0.19 |
| | F | 54 | 2 | 0.09 |
| | Total | 111 | 5 | 0.14 |
| Scargill Valley | M | 37 | 5 | 1.40 |
| | F | 39 | 8 | 0.65 |
| | Total | 76 | 13 | 1.02 |
| Cape Palliser | M | 15 | 11 | 2.69 |
| | F | 23 | 10 | 1.24 |
| | Total | 38 | 21 | 1.97 |
| Castlepoint | M | 27 | 21 | 7.90 |
| | F | 21 | 10 | 3.65 |
| | Total | 48 | 31 | 5.77 |
| Awatere Valley | M | 24 | 16 | 4.64 |
| | F | 22 | 12 | 2.15 |
| | Total | 46 | 28 | 3.40 |

or unwillingly get trapped and there is surely heterogeneity in individual trapability. Are the sample data "representative" to allow a valid inductive inference to the population of interest? I suspect so; however, the authors should ideally make this argument.

## 2.1.3   What Constitutes a "Data Set"?

There is sometimes confusion as to what represents a "data set" in the literature (e.g., Stephens et al. 2005). There are few restrictions on a data set as long as its components have information on the same issue of interest.

Questions concerning the extent of a data set can often be answered by examination of the response variable. In a treatment-control experimental setting, the response variable might be a concentration level of a compound in a blood sample; thus, one data set because both data sets have information on the same issue of interest. However, if different response variables are used (different issues of interest) across some categories, then these constitute different data sets.

I will provide some examples that might help people understand this matter. Consider a simple treatment and control study; one might think there are two "data sets" here, one for the control and another for treatment. Not so – this is to be treated as a single data set. Consider a discriminant function analysis with seven discriminator variables with the analysis being to find out which subset of the seven might serve in a parsimonious model for inference about the discrimination. Here the "data set" consists of the binary response variable and the seven discriminator variables. Of course, given several models in the set, only one (at most, assuming a global model) will have all seven variables in it. Other models will have fewer than seven variables, but this does not invalidate the notion of the "data set" (see Lukacs et al. 2007).

## 2.2 Models

Quantification is nearly essential in the empirical sciences where stochasticity is substantial, there are several different sources of variability (factors), or there is some degree of complexity. This complexity might arise from multiple variables, interactions between and among variables, high variability, nonlinearities (e.g., threshold effects, asymptotes), and a host of other issues. Unless one is engaged in simple descriptive studies, they must deal with mathematical models. Such is certainly the case I focus on here – model based inference. We should not think of this requirement as negative; instead, quantification allows both rigor and the ability to better understand far deeper science issues. Soule (1987:179) offered, "Models are tools for thinkers, not crutches for the thoughtless." Box (1978:436) records that R. A. Fisher felt some statisticians were trained strictly mathematically and that many of them seem to have no experience of the valuable process known as "stopping to think."

We are not trying to model the data; instead, we are trying to model the information in the data. The goal is to recover the information that applies more generally to the process, not just to the particular data set. If we were merely trying to model the data well, we could fit high order Fourier series terms or polynomial terms until the fit is perfect. Data contain both information and noise; fitting the data perfectly would include modeling the noise and this is counter to our science objective. Overfitting is a poor strategy and it goes against the notion of parsimony, a subject to be addressed shortly. Models are *central* to science as they allow a rigorous treatment and integration of:

- Science hypotheses (the all important set $\{H_i\}$)
- Data (e.g., continuous or discrete or categorical),
- Statistical assumptions (e.g., Weibull errors, linearity)
- Estimates of unknown model parameters ($\theta$) and their covariance matrix $\Sigma$

Models are only approximations to full reality. Box (1979) said "… all models are wrong, some are useful." We should think of the value of alternative models as better or worse, instead of right or wrong. While a driver's license is "valid" or not, models do not share this property. The strength of evidence for competing models is very much central to both science and this textbook.

Models must be derived to carefully represent each of the science hypotheses. These models are always to be probability distributions. The idea is that each hypothesis has a model that fully represents it; then we can think of hypothesis $i$ and model $i$ as almost synonyms. That is, the goal is to have a one-to-one mapping between the $i$th hypothesis and the $i$th model:

$$H_1 \Leftrightarrow g_1, H_2 \Leftrightarrow g_2, ..., H_R \Leftrightarrow g_R.$$

People in the life sciences are often poorly trained in modeling techniques; this might be a place where the investigator will want to seek advice or collaboration with a person in the statistical sciences.

Then the science question asks, "What is the support or empirical evidence for the *i*th hypothesis (via its corresponding model), *relative to others in the set.* This leads us to the "model selection" problem. So, finally the issue becomes the *evidence* for each of the hypotheses (and their associated models), *given the data.* Of course, hypotheses and their corresponding models not in the set are out of consideration until, perhaps, they are added at a latter time as the set evolves. So, now we can ask if hypothesis C is 10 times as *likely* as hypothesis A? Is the support for hypotheses A and B nearly equal? Is hypothesis A 655 times more *likely* than hypothesis D? If so, would we take this as very strong evidence? These are the types of science issues that can be answered easily using the existing theory for model based inference.

Often inferences are based on the best hypothesis in the set. While "best" is not defined until Chap. 3, standard analysis often tries to determine or estimate which of the hypothesis is the best, based on the data. Inference is then based on this hypothesis, via its corresponding model – *model based inference.* Assuming models have been derived to represent the hypotheses in the set, this is the so-called model selection problem. A "good" model is able to properly separate information in the data from "noise" or noninformation. Finding such a model is a generic goal of model selection. Now I begin to use the concept that a science hypothesis and its model are (ideally) synonymous.

Many standard approaches to model selection have been developed, including adjusted $R^2$; Mallows' $C_p$; step-up, step-back, and stepwise regression, to name a few. As one might expect, the early approaches are rarely the best ones; what is not expected is that the early methods are still being taught in mainstream statistics classes (at least for nonstatistics majors) and readily available in the most well-known statistical computing packages. Most selection approaches (e.g., stepwise regression) are based on some sort of theory but they are often not based on any underlying theory concerning what is a good *fitted* model, given the data; hence, no rigorous criterion of "best" model. The methods do not have a proper underlying theory, just a semblance of semirelevant theory. The model (or hypothesis) selection issue is central to data analysis: "Which hypothesis/model should I use for the analysis of a particular data set?" and "How can this be best done?"

Approaches are needed to provide quantitative *evidence* for the hypotheses in the set. As information can be quantified in various ways, approaches have been recently developed to address the model selection problem as well as an empirical ranking of the hypotheses in the set, through the associated models. Here it starts to become clear that the modeling step is nearly as important as the hypothesizing step in empirical science.

## 2.2.1   True Models (An Oxymoron)

Models are never "true"; models do not reflect reality in its entirety. In the real world with real data, there is no valid concept of a model that is exactly true, representing full reality. Models are approximations by definition if nothing else. If we had a true model, we would still have to estimate its many parameters and try to interpret the complex result. Any such true model would be quite complicated and involve a great many parameters. Thus, an extraordinarily large sample size would be required, unless it is also assumed that the true model somehow came with its true parameters known to the investigator! I find it hard to imagine a situation where the researcher knew the exact functional form of the true model *and* all of its parameter values! Some scientists might take the view that any such "true" model must be considered infinite dimensioned; perhaps, this is a useful concept but it is just another way of saying there is no valid notion of a true model. Recently I have seen the term "inexact models" used; I believe all models are approximations and, therefore, "inexact."

Computer simulation studies often use Monte Carlo methods to simulate "pseudodata" from a mathematical model, with parameters known or given. Here the exact form of the model and its parameters are known – this is properly termed a generating model. In this computer sense, the "true model" and its parameters are known. A common mistake in the statistical literature has been to provide many replicate pseudodata sets from a generating model, include this model in the set, and then proceed to ask questions about which model selection method most often selects the generating model. Such circular results are of little use in the real world where data arise from complex (and only partially observable) reality, not from a simple parametric model. Real data do not come from models and selection criteria that are designed to select a so-called true model are misguided.

Going further, there is the notion of "consistency" in model selection. Here, some procedures are classed as being "consistent," meaning that as sample size increases (often by three to five orders of magnitude) the probability of selecting the "true" model approaches 1, given the true model is in the set (see Appendix D). This concept seems strained if either the true model is not in the set or if the "true" model is infinite dimensional. In reality, the model set changes as sample size is increased by orders of magnitude and this makes the notion of consistency strained.

The concept of truth and the false concept of a true model are deep and surprisingly important. Often, in the literature, one sees the words *correct* model or simply *the* model as if to be vague as to the exact meaning intended. Bayesians seem to say little about the subject, even as to the exact meaning of the prior probabilities on models. Consider the simple model of population size ($n$) at time $t$

$$n_{t+1} = n_t \cdot s_{t_t},$$

where $s_t$ is the survival probability during the interval from $t$ to $t + 1$. This is a "correct" model in the sense that it is algebraically and deterministically correct; however, it is not an exact representation or model of truth. This model is not explanatory; it is definitional (it is a tautology as it implies that $s_t = n_{t+1}/n_t$). For example, from the theory of natural selection, the survival probability differs among the $n$ animals. Perhaps the model above could be improved if average population survival probability was a random variable from a beta distribution; still this is far from modeling full reality or truth, even in this very simple setting. Individual variation in survival could be caused by biotic and abiotic variables in the environment. Thus, a more exact model of full reality would have, at the very least, the survival of each individual as a nonlinear function of a large number of environmental variables and their interaction terms. Even in this simple case, it is surely clear that one cannot expect any mathematical model to represent full reality – there are no true models in life sciences. We will take a set of approximating models $g_i$, without pretending that one represents full reality and is therefore "true."

Approximating models share some features with maps. Maps fail to capture every detail on the landscape, regardless of their scale. Both data and maps contain errors of omission; this seems unavoidable. Errors of commission should, in principle, be avoided. A map should not show a road or stream that does not exist, while we should not find an effect in the data that does not exist (a spurious effect). All maps are wrong, but some are useful, at least in certain contexts. A map of Switzerland is of limited use in the United States, but might be very useful in Switzerland. Of course, there is no true map.

## 2.2.2  The Concept of Model Parameters

In many cases, parameters are real entities. For example, the size of a population of parrots in an aviary can be determined by a census at a given point in time and this count is a parameter ($N$, the population size). If we have a time series of censuses of this population, the parameters are $N_t$, where $t$ is time. However, parameters are often human constructs and are important in understanding systems or processes. The probability of death in a fish population in a large lake is unobservable and not really a parameter in some sense. Instead, we, as investigators, define an arbitrary time interval (such as a month or a year) and derive models that include the probability of death as a parameter, and proceed to estimate this. It is a model parameter, but is it a parameter intimately associated with the population of fish? Perhaps not? Linear regression models are merely first-order approximations to often complex processes of interest. Any particular $\beta$ (regression "slope") is unlikely to be a parameter associated with the process itself. Similarly, $\lambda$, the finite rate of population change, is hardly a parameter that can be directly observed or measured, but it serves as a very useful construct in population ecology. Scientific understanding can often be aided by the notion of parameters, whether real or just useful or directly observed or unobservable.

In fact, thinking that truth is parameterized is itself a type of (artificial) model based conceptualization. Going deeper, mathematics itself is a "model" when used to represent reality or concepts or hypotheses. Mathematics is a human construct and does not exist in the same sense as reality. Sometimes it is useful to think of $f$ as full reality and let it have (conceptually) an infinite number of parameters. This "crutch" of infinite dimensionality at least keeps the concept of reality even though it is in some unattainable perspective. Thus, $f(x)$ represents full truth, and might be conceptually based on a very large number of parameters (of a type we have not even properly conceived) that gives rise to a set of data $x$.

Akaike noted that the success of the analysis of real data depends essentially on the choice of the basic model. Successful use of statistical methods depends on the integration of subject-matter science into the statistical formulation. This demands a significant amount of effort for each new problem. This is where the science of the issue enters consideration: a major step.

### 2.2.3   Parameter Estimation

It is a *fitted* model that is the basis for statistical inference; hence, parameter estimation is very important. If the sample size is small, the parameter estimates will typically have large variances and wide confidence intervals and might be so uncertain as to be of little use. Large sample size conveys many important advantages in terms of parameter estimates and model selection.

Given a model and relevant data, procedures were developed nearly a century ago to estimate model parameters. Three common approaches have emerged for general parameter estimation: least squares, LS (or "regression"), maximum likelihood, ML, and Bayesian methods. Least squares has been popular; however, its domain is primarily the class of the so-called general linear models (e.g., regression and ANOVA). I will say little about this approach. The much more general approach is Fisher's maximum likelihood (see Appendix A). The notion of ML is compelling – given a model and data, taking as the estimate the value of the parameter that is "most likely." Hence the name maximum likelihood estimate (MLE); it is the value of the parameter that is most likely, given the data and model.

As sample size increases (asymptotically), MLEs enjoy several properties (within certain regularity conditions): unbiased, minimum variance, and normally distributed. In addition, if one takes an MLE and transforms it to another estimate, it too is an MLE (the "invariance" property). These are important properties and explain partially why likelihood is so central to statistical thinking.

An important component of data analysis relates to the fit of a model to the data. These activities focus primarily on a global model and include such things as a formal goodness-of-fit test, adjusted $R^2$ value, residual analyses, and checking for overdispersion in count data. If the global model is judged to be "poor," then further data analysis will likely be compromised.

A person new to statistical thinking often finds it difficult to relate data, model, and model parameters that must be estimated. These are hard concepts to understand and the concepts are wound into the issue of parsimony. Let the data be fixed and then realize the information in the data is also fixed, then some of this information is "expended" each time a parameter is estimated. Thus, the data will only "support" a certain number of estimates, as this limit is exceeded parameter estimates become either very uncertain (e.g., large standard errors) or reach the point where they are not estimable.

### 2.2.4  Principle of Parsimony

A model has structural and residual components. Parsimony relates to under- and overfitting models. Examination of the graph in Fig. 2.1 shows that an underfitted model (the left side of Fig. 2.1) risks not only high bias, but also the illusion of high precision ("a highly precise wrong answer"). Underfitting relates to the case where some model structure is erroneously included in the residuals. Of course the investigator does not necessarily know the situation she is in. Overfitted models are also to be avoided because further examination of the graph suggests that overfitting (the right side of Fig. 2.1) risks including too many parameters (that need to be estimated) and a high level of uncertainty. Overfitting relates to the case where some residual variation is included as if it were structural. This may seem like the lesser of two evils;

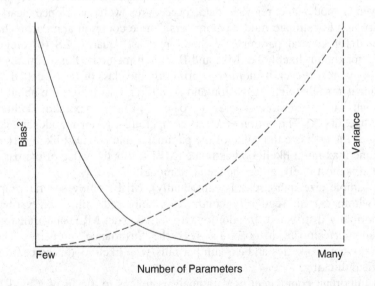

FIG. 2.1. The Principle of Parsimony is illustrated here as a function of the number of esti-mable parameters ($K$) in a model. There are two processes here: first, bias (or squared bias) declines as $K$ increases and, second, the variance (uncertainty) increases as $K$ increases. These concepts suggest a trade-off whereby the effects of underfitting and overfitting are well balanced.

however, precision is lessened, often substantially. Overfitting implies that some noise (noninformation) has been included in the structural part of the model and the effects are not part of the actual process under study (i.e., spurious). Edwards (2001:129) says it in an interesting way,

*"...too few parameters and the model will be so unrealistic as to make prediction unreliable, but too many parameters and the model will be so specific to the particular data set so to make prediction unreliable."*

Clearly, one wants a proper trade-off between squared bias vs. variance or, said another way, between under- and overfitting. Either extreme will result in unreliable prediction. Residuals might be pure noise or information that cannot be decoded yet. The concepts of under- and overfitting depend on sample size; as sample size increases, additional information is available in the data, and smaller effects can be identified. Thus, residual variation can be understood and this transfers to the structural part of the model. Parsimony cannot be judged against any notion of a true model.

The concept of parsimony in modeling and estimation has been an important statistical principle for several decades. The general notion of parsimony has a much longer history in science and engineering and is closely related to Occam's razor. Parsimony is a fundamental issue in science and it is easy to overlook its depth and importance. Occam's statement has a literal translation from Latin, but is commonly referred to as "Occam's razor" meaning roughly to "shave away all that is not needed."

Parsimony appears to be a simple notion; however, it is easy to underrate its importance and its centrality in modeling, model selection, and statistical inference. Parsimony can be viewed as a trade-off between squared bias and variance (variance is a squared quantity, thus bias is squared for some comparability). Think of parsimony as a function of the number of estimable parameters in a model (denote this parameter count $K$). Given a fixed data set, two things happen as the number of model parameters to be estimated are increased (the standard example is a polynomial where additional parameters are introduced from a linear, to a quadratic, to a cubic, etc.). First, squared bias decreases as more parameters are added – this is good. Second, uncertainty (measured by the variance) increases as more parameters are added – this is not so good (Fig. 2.1).

The addition of more parameters reduces bias but, in doing so, increases the uncertainty. That is, for a given data set and its context, there is a "penalty" or "cost" for adding more parameters that must be estimated. It is the need to *estimate* parameters from the data that is the difficulty. If one could somehow add parameters with *known* values, the situation would be simple: that is, consider only models with a large number of parameters. Unfortunately, parameters in these models are not known; reality is harsh in this regard and parameters must be estimated based on the information in the data. Each time a parameter is estimated, some information is "taken out" of the data, leaving less information available for the estimation of still more parameters.

Parsimony exists near the small region where the lines cross – a trade-off (Fig. 2.1). Parsimony is a conceptual goal because neither bias nor variance is known to the investigator analyzing real data. There are many specific approaches to achieving parsimony but the important concept does not, by itself, lead to a specific criterion or recipe. Parsimony is a property of models (and their parameters that must be estimated) and the data.

There is a large literature admonishing investigators to avoid overfitting as this leads to spurious effects and imprecision. An equally large literature warns of underfitting because of bias and effects that are present, but missed during data analysis. Until somewhat recently, statistical science lacked an effective way to objectively judge the trade-off – how many are too many, how many are not enough. This has been largely resolved for a wide class of problems and is another example of the advantage (actually necessity) of quantification. Rigor in empirical science has a basis in quantification. All methods for model selection are linked in some manner with the principle of parsimony.

I have had biologists state that "A biologically reasonable model is 'punished' because it has too many unknown parameters." Indeed, the estimation of parameters sucks information from the data to the point that little or no information is left for the estimation of still more parameters. It is easy, at first, to think that parameters come somehow "free" and that complex biological models can be developed with little or no data. Instead, the reality of the situation is that parameter uncertainty must harken back to the concept of parsimony. A partial solution to obtaining increased biological reality is to obtain a large sample size or improve study design (e.g., control some factors) as these allow parameter estimates with good precision and functional model forms to be evaluated.

In model selection, we are really asking which is the best model *for a given sample size*. Given a real process that has some realistic degree of complexity and high dimensionality, a high-dimensioned model might be selected as best if the sample was quite large. In the same situation, a small, low-dimensioned model might be expected if the sample was small. A very rough rule of thumb advises that at most $n/10$ parameters can be estimated; thus for observations on a sample of 30 individuals, one might be able to estimate about three parameters (e.g., $\beta_0$, $\beta_1$, and $\sigma^2$) in a regression model. This is often less than what biologists attempt with such small data sets.

Model selection resulting from the analysis of sparse data usually suggest a simple model with few parameters. Such results should not be taken to suggest that the system under study is necessarily simple. On the contrary, if a virtually "null" model is selected, this usually points to an insufficient amount of data to fit anything more realistic. Even then, if the best model is, for example, one with no time effects, one should not infer the process is time invariant. Instead, the correct interpretation is that the variation in some parameter across time is small and such variation could not be identified with the small amount of information in the data.

We are really asking – how much model structure will the data support? A good fit is *not* sufficient, we need predictive ability, and this involves parsimony – how many parameters can be estimated and included in a model? Overfitting risks (by the addition of extra parameters) the inclusion of some of the random "noise" as if it were structure. Model selection criteria allow an objective measure of how many parameters can be fitted to a model, given the sample size. We can chase truth, but we will never catch it and parsimony is central to the chase.

### 2.2.5  Tapering Effect Sizes

In perhaps all of the empirical sciences, there are a wide range of "effect sizes." There are the large, dominant effects that can often be picked up even with fairly small sample sizes and fairly poor analytical approaches (e.g., stepwise regression). Then there are the moderate-sized effects that are often unveiled with decent sample sizes and more adequate analysis methods. It is more challenging to identify the still smaller effects: second- and third-order interactions and slight nonlinearities. Increasingly large samples are needed to reliably detect these smaller effects. Beyond these small effects lie a huge number of even smaller effects or perhaps important effects that stem from rare events. This situation is common as any field biologist can attest. We say there are "tapering effect sizes" and we can chase these with larger sample sizes, better study design, and better models based on better hypotheses. The notion of tapering effect sizes is everywhere in the real world and it is hard to properly emphasize their importance.

Tapering effect sizes are what preclude the notion of a true model. Just the high-order interactions are quite complex. Consider the ramifications of the various systems in the human body as body temperature climbs to 105° or as one finishes a marathon run. The life sciences are all about a wide variety of tapering effect sizes.

## 2.3   Case Studies

### 2.3.1  Models of Hardening of Portland Cement Data

This is a well-known data set and authors typically approach the issue as a multiple linear regression problem with four predictor variables. The global model is

$$E(y) = \beta_0 + \beta_1(x_1) + \beta_2(x_2) + \beta_3(x_3) + \beta_4(x_4),$$

where $y$ is the calories of heat evolved per gram of cement after 180 days, $x_1$ the percent calcium aluminate ($3\text{CaO} \cdot \text{Al}_2\text{O}_3$), $x_2$ the percent tricalcium silicate ($3\text{CaO} \cdot \text{SiO}_2$), $x_3$ the percent tetracalcium alumino ferrite ($4\text{CaO} \cdot \text{Al}_2\text{O}_3 \cdot \text{Fe}_2\text{O}_3$),

$x_4$ the percent dicalcium silicate ($2CaO \cdot SiO_2$), and $E(\cdot)$ is the expectation operator (see Appendix B).

This example problem has two objectives: variable selection and prediction. The analysis could be done in a least squares (LS) or maximum likelihood (ML) framework (Appendix A). The LS and ML estimates of the $\beta_i$ parameters will be identical; the two estimates of $\sigma^2$ will differ slightly. This might be a place for the reader to review quantities such as the residual variance $\sigma^2$, residual sum of squares RSS, adjusted $R^2$, the covariance matrix $\Sigma.$, various residual analyses, and the notion of a global model.

Because only four variables are available, the temptation is to consider all possible models ($2^4-1 = 15$) involving at least one of the regressor variables. Burnham and Anderson (2002), strictly as an exploratory example, considered the full set of models, including the global model {1234} with $K = 6$ parameters (i.e., $\beta_0$, $\beta_1$, $\beta_2$, $\beta_3$, $\beta_4$, and $\sigma^2$). They generally advise against consideration of all possible models (15 in this example) of the $x_i$ (note that even more models would be needed if interactions, powers of the predictor variables, or other nonlinear relationships were employed).

In contrast, for this example, I will try to limit the set to those that seem plausible, particularly in view of the small sample size. Using all possible models usually represents an unthinking, naive approach. I have already noted that the global model is essentially singular as the numerical values for the four variables sum to approximately 1 (rounding prevents some sums to be exactly 1). Thus, the global model can be dismissed in the example. I already eliminated the four single variable models as cement is a mixture of ingredients. So, now the set is down to 10 models.

Additional thinking (Sect. 1.8) about the chemical similarity of the pair of variables 1 and 3 and the pair 2 and 4 was relevant. Without the curse of data dredging, it is advisable to examine the correlations between these pairs, based on the data available. Such correlation analysis substantiates the observation; the correlation coefficient between $x_1$ and $x_3$ was $-0.824$ and the correlation between $x_2$ and $x_4$ was $-0.973$). Thus, including both variables within a pair would not be advisable, particularly in view of the fact that the sample size is only 13 observations. However, we do not know if $x_1$ or $x_3$ is the better predictor, nor do we know if $x_2$ or $x_4$ is the better predictor. Thus, the following five hypotheses and variables lead to five models making up the candidate set:

| $H_1$ | 0 variables | $g_1$ | $E(y) = \beta_0$ |
|---|---|---|---|
| $H_2$ | $x_1$ and $x_2$ | $g_2$ | $E(y) = \beta_0 + \beta_1(x_1) + \beta_2(x_2)$ |
| $H_3$ | $x_1$ and $x_2$ and $x_1 * x_2$ | $g_3$ | $E(y) = \beta_0 + \beta_1(x_1) + \beta_2(x_2) + \beta_3(x_1 * x_2)$ |
| $H_4$ | $x_3$ and $x_4$ | $g_4$ | $E(y) = \beta_0 + \beta_1(x_3) + \beta_2(x_4)$ |
| $H_5$ | $x_3$ and $x_4$ and $x_3 * x_4$ | $g_5$ | $E(y) = \beta_0 + \beta_1(x_3) + \beta_2(x_4) + \beta_3(x_3 * x_4)$ |

Hypothesis $H_1$ has no predictor variables and is not in the original 15 possible models. I include it here as an example. Of course, the numerical values for the ML estimates of the $\beta$ parameters will differ across models (i.e., $\beta_1$ "means" different things and is model specific). Now it becomes clear that hypothesis 4 ($H_4$), for example, has its corresponding model ($g_4$). This is a set of first-order models with all the variables entering a linear model. The model set is crude; however, there are little data and so more complex models might not be justified. Note how knowledge of sample size affects the number of parameters that might reasonably be estimated; this requires some experience. However, even a C student just finishing a class in applied regression would surely not attempt to estimate 8–10 parameters from this data set. This will serve as our initial example in later chapters.

The cement data have high levels of dependencies (correlations) among the predictor variables as is typical of most problems where a regression analysis might be appropriate. If all the regressor variables are mutually orthogonal (uncorrelated) then analytical considerations are more simple. Orthognality arises in controlled experiments where the factors and levels are *designed* to be orthogonal. In observational studies, there is often a high probability that some of the regressor variables will be mutually quite dependent. Rigorous experimental methods were just being developed during the time these data were taken (about 1930). Had such design methods been widely available and the importance of replication understood, then it would have been possible to break the unwanted correlations among the $x$ variables and establish cause and effect if that was a goal.

## 2.3.2 Models of Bovine TB Transmission in Ferrets

Caley and Hone's (2002) models for disease transmission dealt with the age-specific force of infection, $\lambda(a)$ for various age classes and the age-specific disease prevalence model with ($\alpha > 0$) and without ($\alpha = 0$) disease-induced mortality. Their model for $H_1$ without disease-induced mortality was

$$1 - e^{-\lambda a},$$

where $\alpha \leq s$ and $s$ is the suckling period. The corresponding model for $H_1$ with disease-induced mortality was

$$\frac{\lambda \left(1 - e^{(\alpha - \lambda)a}\right)}{\lambda - \alpha e^{(\alpha - \lambda)a}}.$$

Simple graphs of their hazard rates help in understanding the models derived. They introduced a guarantee parameter ($\mathcal{I}$) for the period when ferrets were not exposed to infection (Fig. 2.2). If $a > s$, then the corresponding hazard models are

$$1 - e^{-\lambda s},$$

where $s$ is the suckling period and no disease-induced mortality, and

$$\frac{\lambda (1 - e^{(\alpha - \lambda)a})}{\lambda - \alpha e^{(\alpha - \lambda)a}} e^{-\alpha(a - s)}$$

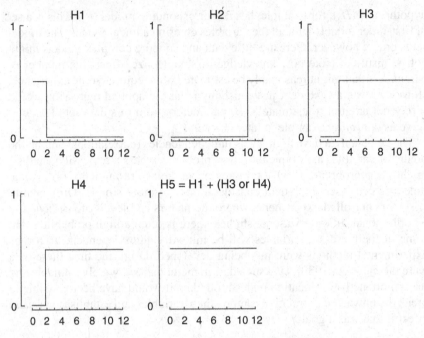

FIG. 2.2. Constant hazard functions used by Caley and Hone (2002) in modeling hypotheses concerning tuberculosis transmission in feral ferrets in New Zealand.

with disease-induced mortality. In addition, all models had a gender effect and a site effect (data were collected at seven sites). The hazard models become tedious (see their Table 2.1) and they then defined $p_i$ to be the modeled probability of infection. A binomial likelihood function was then used where the data were $y$ = the number of infected individuals from a total of $n_i$ in each gender class (see data in Table 2.2). Thus, $\hat{p}_i = y_i / n_i$ as an estimate of $p = E(y_i)/n_i$. Several bounds and constraints were placed on parameter values during the optimization; additional details are given by Caley and Hone (2002). Clearly, a great deal of effort was made to derive models that accurately portrayed the hypotheses about disease transmission and the force of infection ($\lambda$) as functions of age, gender, guarantee time, and disease-induced mortality.

## 2.4    Additional Examples of Modeling

The first example provides some details on models and how the models might help in developing interesting science hypotheses. This is followed by further considerations in the *Exercises* section. The other two examples are more typical where the task is merely to well represent the science hypotheses by models.

## 2.4.1 Modeling Beak Lengths

Beak size bimodality in Darwin's finches (*Geospiza fortis*) on the island of Santa Cruz, Galapagos, Ecuador has been of interest since the early 1960s. Hendry et al. (2006) provide some background and analysis results on this set of evolutionary issues. Here we will take a hypothetical view of the data and general science question and provide alternative approaches to provide insights into hypothesizing and modeling. This example will use just beak length, while Hendry et al. (2006) performed a principal components analysis on several measurements to estimate beak "size." I will not address these real world complexities here as I want to focus on a different way to approach the evolutionary questions of interest. This approach is not claimed to be better in any way; only different to give the reader a feeling for both hypothesizing and modeling. Interested readers are encouraged to read Hendry et al. (2006) for their results with the real data.

Beak length data were collected on 1,755 birds during 1964–2005 at Academy Bay, adjacent to the town of Puerto Ayora. Histograms of the measurements suggested bimodality in the early years; however, this bimodality was lost in concert with marked increases in human population density and activity over time. This observation led to hypotheses about evolutionary forces promoting bimodality and driving adaptive radiation into multiple species over time. Perhaps the increased human disturbance blocked or at least hampered the radiation and bimodality in recent years at this site. While this extension of the problem is only to illustrate some principles, it will follow some aspects of the real situation described by Hendry et al. (2006).

Before proceeding, it is interesting to note a confirmatory aspect of this study. There are many variables that have changed on this island over the past 40–50 years. A descriptive approach might have taken measurements on many variables and asked which is the better predictor or which variables have the highest adjusted $R^2$ value? This is a "shot gun approach" and exposes the investigator to a high probability of finding spurious effects. The confirmatory approach asks a more specific question (is human disturbance associated with evolutionary changes in bill length?), after trying to think hard about the issue.

We first hypothesize that the bimodality observed in the histograms (per year sample sizes were roughly 100) was largely an artifact. Perhaps another bin size for the histograms would not show any pronounced bimodality. Thus, we begin with the hypothesis ($H_1$) that the sample data were taken from a unimodal population (this model might be of particular use in the analysis of data for the later years). Beak lengths cannot be negative; so I will use the gamma model instead of the more usual normal model. The gamma distribution (denote this first, unimodal, model as $g_1$) is

$$g_1(x) = \frac{x^{\alpha-1} e^{-x/\beta}}{\Gamma(\alpha) \beta^\alpha}.$$

This model (a PDF) has two parameters, $\alpha$ and $\beta$, and $x$ is the beak length. Note the expanded model notation to make it clear that the model is a function of the data $x$. This complicated looking model has a nice simple form (Fig. 2.3) and seems adequate as a mathematical representation of the measurement data on beak lengths (assuming unimodality).

This distribution is useful in that its mean value is estimated as $\hat{\alpha}\hat{\beta}$ and the variance is estimated as $\hat{\alpha}\hat{\beta}^2$. The shape of the distribution changes depending on the values of $\alpha$ and $\beta$; thus, the gamma distribution, like many statistical distributions, is a family of curves. Note too, in this type of modeling there is no response variable, instead it is the distribution of bill lengths ($x$) that is being modeled.

Considering the apparent observed bimodality, one might consider a *mixture* of two gamma distributions as a second model. This hypothesis assumes there is a small-beaked phenotype with some variability across individuals around a mean. Similarly, a large-beaked phenotype has some variability across individuals around a different (larger) mean. Thus, it is quite possible that an individual from the small-beaked phenotype might have a longer beak than an individual from a large-beaked phenotype. The data are hypothesized to be an unknown mixture of the two (perhaps highly variable) phenotypes. These considerations lead to a model ($g_2$) for the second hypothesis, $H_2$ (Fig. 2.4):

$$g_2 = \pi(g_s) + (1-\pi)(g_b),$$

where the parameter $\pi$ is the "mixture coefficient" and $g_s$ and $g_b$ are gamma distributions for small (s)- and big (b)-beaked individuals, respectively. Here, $0 \le \pi \le 1$ and is the proportion of the population that have small beaks. For example, if 17% of the individuals in a given year were from a population of small-beaked phenotypes, then

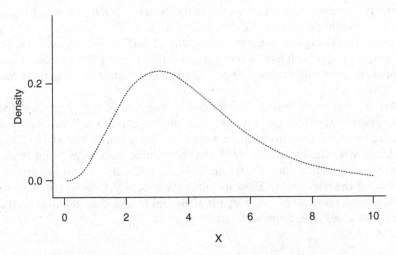

FIG. 2.3. The hypothesis of unimodality in finch bill lengths is represented as a gamma distribution.

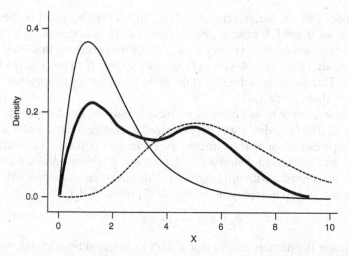

FIG. 2.4. The model representing the hypothesis that finch bill lengths arise from an unknown mixture of two phenotypes, each phenotype is modeled as a gamma distribution.

$$g_2 = 0.17(g_s) + (0.83)(g_b).$$

Of course, the two gamma distributions above would each have parameters $\alpha$ and $\beta$ to specify the exact shape of the distributions. This model has five parameters: $\pi$, $\alpha$, and $\beta$ for the small-beaked animals and an $\alpha$ and $\beta$ for the big-beaked animals. If we have a way to measure the strength of evidence for these two models we could answer questions about unimodality vs. bimodality: i.e., compare models $g_1$ vs. $g_2$.

Now we hypothesize ($H_3$), a linear change in bimodality over years and this is easily done by adding a submodel on the mixture coefficient $\pi$. We take model 2 and extend it to obtain a model that allows bimodality to change (drift) over years:

$$g_3 = \pi(g_s) + (1 - \pi)(g_b)$$

and replace the parameter $\pi$ (in two places) with the submodel

$$\pi = \beta_0 + \beta_1(T),$$

where $T$ is the year of the study. The parameter $\pi$ no longer appears in the model as it is replaced by the submodel that allows the mixture to be a function of year. Carrying out this substitution,

$$g_3 = \left(\beta_0 + \beta_1(T)\right) \cdot (g_s) + \left(1 - (\beta_0 + \beta_1(T))\right) \cdot (g_b).$$

This model has six parameters: $\beta_0$, $\beta_1$, $\alpha$, and $\beta$ for the small-beaked animals, and an $\alpha$ and $\beta$ for the big-beaked animals ($\pi$ has been deleted and the two $\beta$ parameters added). Hendry et al. (2006) hypothesized that bimodality decreased after the first few years ($T$), thus we expect $\hat{\beta_1}$ to be negative in this example. This model gets directly at the main evolutionary hypotheses; this is the role of these models.

We can hypothesize still other plausible alternatives, as Chamberlin would have urged. The bimodality was hypothesized to change over years (that is $H_3$) but perhaps caused or at least influenced by human population density over time ($T$) and associated disturbance (denote this environmental covariate as $X_1$ not to be confused with bill length, $x$). This covariate was measured; so we have the model ($g_4$) to represent the fourth hypothesis ($H_4$):

$$g_4 = \pi(g_s) + (1 - \pi)(g_b).$$

Now replace the mixture coefficient $\pi$ with a similar submodel, but with the human covariate as

$$\pi = \beta_0 + \beta_1(X_1).$$

In a sense, $g_3$ asked *what?* while $g_4$ begins to ask *why?*

We now turn our attention to a supposed covariate dealing with yearly precipitation (denote this as $X_2$). We will assume this variable has been measured and we will let it enter the analysis as binary: 1 for heavy precipitation and 0 for virtually no rainfall (the usual case). One can already see the pattern here as we will hypothesize ($H_5$) that the bimodality is influenced by (only) precipitation over the years of the study. Its associated model is

$$g_5 = \pi(g_s) + (1 - \pi)(g_b),$$

where $\pi = \beta_0 + \beta_1(X_2)$.

An astute biologist then hypothesizes ($H_6$) that bimodality is influenced by both human activity ($X_1$) and precipitation ($X_2$), leading to an expanded submodel for $\pi$:

$$g_6 = \pi(g_s) + (1 - \pi)(g_b),$$
$$\pi = \beta_0 + \beta_1(X_1) + \beta_2(X_2).$$

Finally, investigators hypothesize ($H_7$) to reflect interest in an interaction term in the submodel for $\pi$ as

$$g_7 = \pi(g_s) + (1 - \pi)(g_b),$$
$$\pi = \beta_0 + \beta_1(X_1) + \beta_2(X_2) + \beta_3(X_1 * X_2).$$

This model has eight parameters: $\beta_0$, $\beta_1$, $\beta_2$, $\beta_3$, $\alpha$, and $\beta$ for the small-beaked animals, and an $\alpha$ and $\beta$ for the big-beaked animals. [I hope it is clear that the $\beta$ parameters differ from model to model; i.e., the value for the MLE for $\beta_1$ and the interpretation differ by model.]

Given the sample of 1,755 beak measurements and the seven models of the seven science hypotheses, one could estimate the model parameters using maximum likelihood (see Appendix A) and proceed with a formal analysis of the evidence for each of the seven. Note that there is a nice one-to-one mapping of each hypothesis with its model. Of course, each submodel could have been hypothesized to be quadratic (or even cubic), but additional $\beta$ parameters would be needed to chase these potential nonlinearities. This example attempts to show how hypothesizing and modeling can have catalytic effects. We will see this example again in the subsequent chapters.

## 2.4.2 Modeling Dose Response in Flour Beetles

Young and Young (1998:510–514) give as an example (originally from Bliss 1935) of modeling acute mortality of flour beetles (*Tribolium confusu*) caused by an experimental five-hour exposure to gaseous carbon disulfide ($CS_2$). The data are summarized in Table 2.3. The sample size is the 471 beetles in the dose–response experiment. One can see from Table 2.3 that the observed mortality rate increased with dosage. It is typical to fit a parametric model to effectively smooth such data, hence to get a simple estimated dose–response curve and confidence bounds, and to allow predictions (perhaps even outside the dose levels used in the experiment (i.e., extrapolation)).

A generalized linear models approach may easily, and appropriately, be used to model the probability of mortality, $\pi_i$, as a function of dose level $x_i$. The likelihood function for the data for a single dose is assumed to be binomial and is proportional to

$$\mathcal{L}(\pi \mid n \text{ and } y, \text{binomial}) \propto \pi^y (1-\pi)^{n-y}.$$

This notation (above) is read – the likelihood of the unknown mortality parameter $\pi$, *given* the data (the $n$ and $y$) and the binomial model. The likelihood function would be different with different data or when using a model other than the binomial. Use of the binomial model brings certain assumptions with

TABLE 2.3. Flour beetle mortality at eight dose levels of $CS_2$ (from Young and Young 1998).

| Dose (mg/L) | Number of beetles | | Observed mortality rate |
| --- | --- | --- | --- |
| | Tested | Killed | |
| 49.06 | 49 | 6 | 0.12 |
| 52.99 | 60 | 13 | 0.22 |
| 56.91 | 62 | 18 | 0.29 |
| 60.84 | 56 | 28 | 0.50 |
| 64.76 | 63 | 52 | 0.83 |
| 68.69 | 59 | 53 | 0.90 |
| 72.61 | 62 | 61 | 0.98 |
| 76.54 | 60 | 60 | 1.00 |

it, such as independence). Note, as is always the case, likelihoods are products of probabilities and functions of only the unknown parameters; everything else is known (i.e., given). Shorthand notation includes $\mathcal{L}(\pi|\text{data})$ or just $\mathcal{L}$ if the context is clear. The symbol "$\propto$" means "proportional to" because a constant term (the binomial coefficient), independent of the model parameters, has been omitted (Appendix A).

The flour beetles were dosed at eight levels and the likelihood for the entire data set is merely a product of the eight binomial likelihoods (given the usual assumption of independence, which seems quite reasonable here):

$$\mathcal{L}\left(\pi_i \mid n_i \text{ and } y_i, \text{binomial}\right) \propto \prod_{i=1}^{8}\left(\pi_i\right)^{y_i}\left(1-\pi_i\right)^{n_i-y_i}$$

or just the shorthand

$$\mathcal{L}\left(\pi_i \mid \text{data}\right) \propto \text{ or } L \propto \prod_{i=1}^{8}\left(\pi_i\right)^{y_i}\left(1-\pi_i\right)^{n_i-y_i}.$$

This likelihood sets up a model of the unknown mortality probabilities, but they do not depend on dose. Thus, we can hypothesize some monotonic parametric submodels involving dose $\pi_i \equiv \pi(x_i)$. I will denote dose at level $i$ simply as $x_i$ and constrain the probability of mortality ($\pi$) to be within 0–1.

In the context of generalized linear models, there must be a nonlinear transformation (i.e., link function) of $\pi(x)$ to give a linear structural model in the parameters. There are several commonly used forms for such a link-function based linear model but no single model form that is theoretically the correct, let alone true, one. We consider three commonly used generalized linear models and associated link functions: logistic, hazard, and probit. Each of these models has two unknown parameters that may be estimated from the data using ML. The logistic model form is

$$\pi(x) = \frac{1}{1+e^{-(\alpha+\beta x)}}$$

with link function

$$\log\left(\frac{\pi(x)}{1-\pi(x)}\right) = \log\text{it}(\pi(x)) = \alpha + \beta x.$$

The hazard function and the associated complementary log–log link function are

$$\pi(x) = 1 - \exp\{-e^{(\alpha+\beta x)}\}$$

and

$$\log[-\log(1-\pi(x))] = \text{clog}\log(\pi(x)) = \alpha + \beta x.$$

The cumulative normal model and associated probit link are

$$\pi(x) = \int_{-\infty}^{\alpha + \beta x} \left[ \frac{1}{\sqrt{2\pi}} e^{-(1/2)z^2} \right] dz \equiv \Phi(\alpha + \beta x)$$

and

$$\Phi^{-1}(\pi(x)) = \text{probit}(\pi(x)) = \alpha + \beta x.$$

Here, $\phi(\bullet)$ denotes the standard normal cumulative probability distribution, which does not exist in closed form.

In each of the three cases above, the model is sigmoidal, bounded by 0 and 1, and has two parameters. These are little more than descriptive models; i.e., they have about the "right" shape and have been useful in this class of experiments since the 1930s. The link functions let the investigator "think" of the models as simple regressions, $\alpha + \beta x$ and this is a useful construct.

Substituting the logistic model for dose level into the likelihood for a product of binomials gives

$$\mathcal{L} \propto \prod_{i=1}^{8} \left( \frac{1}{1 + e^{-(\alpha + \beta x_i)}} \right)^{y_i} \cdot \left( 1 - \frac{1}{1 + e^{-(\alpha + \beta x_i)}} \right)^{n_i - y_i},$$

where $x_i$ is the dose level and the likelihood ($\mathcal{L}$ above) is formally $\mathcal{L}(\alpha, \beta | \text{data})$. Thus ML can be used to get the MLEs $\hat{\alpha}$ and $\hat{\beta}$ (the probabilities of mortality are removed and they are replaced by a simple function of dose level, $x_i$). This particular example happens to be logistic regression and can be done easily in software packages (e.g., SAS Institute 2004). This is another example where one might start with a simple model, such as the binomial model here. Assuming independence (Sect. 6.1 and Appendix A), one can take the product of all eight binomials as the likelihood. Then, adding submodels in place of one or more model parameters can bring tremendous flexibility and realism to the modeling process.

At some early point we must ask if a model fits the data in a reasonable way. A simple Pearson observed vs. expected chi-square comparison often suffices as a goodness-of-fit (GOF) assessment:

$$\chi^2 = \sum \frac{(O_j - \hat{E}_j)^2}{\hat{E}_j}$$

where $O_j$ is the observed values and $\hat{E}_j$ is the estimated expected values. These test statistics each have six degrees of freedom (=8 − 2, as each model has two estimated parameters). The chi-square statistic is

$$\chi^2 = \sum_{i=1}^{8} \frac{(y_i - n_i \hat{\pi}_i)^2}{n_i \hat{\pi}_i (1 - \hat{\pi}_i)}.$$

Goodness-of-fit results are

| Model | $\chi^2$ |
|---|---|
| cloglog | 3.49 |
| probit | 7.06 |
| logit | 7.65 |

indicating a good fit for each of the three models. Normally GOF is assessed only for the global model; in this case there is no such model, but three non-nested competitors all with $K = 2$ unknown model parameters (see Appendix A). If the response variable is continuous, there are a large number of standard diagnostics and procedures to analyze residuals; these are widely available in computer software.

A key feature of this beetle mortality example is causality. The experimentally applied dose *caused* the observed mortality. By the design we can establish *a priori* that (1) the only predictor needed, or useful, is dose and (2) monotonicity of expected response should be imposed (i.e., the higher the dose, the higher the probability of death). The issue about a model is thus reduced to one of an appropriate functional form, hence, in a generalized linear models framework, to what is the appropriate link function. However, as a result, we have no global model, but rather several (three were used) alternatives for a best causal-predictive model (many observational studies lack a global model).

### 2.4.3   Modeling Enzyme Kinetics

Over many years, a series of models have been developed for understanding enzyme inhibition (Nelder 1991; Brush 1965, 1966). This field has matured and I will give some general models that have found use in this issue. We will review four models, each representing a hypothesis concerning the rate of enzyme-mediated reaction ($R$). There are only two predictor variables: $S$ = substrate concentration and $I$ = inhibiting substance. Four hypotheses are represented by the following models:

$H_1$ noncompetitive (general) model                  Parameters

$$R = \frac{\beta_1 \cdot S}{\beta_2(1 - \beta_3 \cdot I) + S(1 + \beta_4 \cdot I)}$$         $\{\beta_1,\beta_2,\beta_3,\beta_4,\sigma^2\} = 5$

$H_2$ Michaelis–Menten model

$$R = \frac{\beta_1 \cdot S}{\beta_2 + S}$$         $\{\beta_1,\beta_2,0,0,\sigma^2\} = 3$

$H_3$ competition model

$$R = \frac{\beta_1 \cdot S}{\beta_2(1 - \beta_3 \cdot I) + S} \qquad\qquad \{\beta_1, \beta_2, \beta_3, 0, \sigma^2\} = 4$$

$H_4$ uncompetitive model

$$R = \frac{\beta_1 \cdot S}{\beta_2 + S(1 + \beta_4 \cdot I)} \qquad\qquad \{\beta_1, \beta_2, 0, \beta_4, \sigma^2\} = 4$$

Two of the model parameters have scientific interpretations: $\beta_1$ is the maximum reaction rate and $\beta_2$ is the half saturation level. Parameters $\beta_3$ and $\beta_4$ are called inhibition kinetic values. Here each hypothesis has been represented by its associated model and we can speak of hypothesis $i$ or model $i$ synonymously. The parameters can be estimated using ML methods and inferences made. Critically, we would like measures of the evidence for each of the four hypotheses in the set: "What is the empirical support for hypothesis $i$ vs. $j$?" A set of models such as this does not arise overnight; instead, these models are the result of much effort in the laboratory and much analytical thought. Model building should take full advantage of past research.

There are many model based studies in human medicine but my opinion is that often only a single hypothesis and its model are the focus of the study. Clyde (2000) and Remontet et al. (2006) provide examples of multiple models and model selection in this important area.

## 2.5  Data Dredging

Data dredging (also called *post hoc* data analysis) begins after the planned (*a priori*) analysis and after inspecting those results. Data dredging should generally be minimized or avoided, except in (1) the early stages of exploratory work or (2) *after* a more confirmatory analysis has been completed. In this latter case, the investigator should fully admit to the process that lead to the *post hoc* results and should treat the results much more cautiously than those found under the initial, *a priori* approach. One approach in *post hoc* analyses is to start with the best model (from the *a priori* results) and expand around it. When done carefully, we encourage people to explore their data beyond the important *a priori* phase. Still, *post hoc* results are like skating on thin ice – lots of risks of getting in trouble (i.e., finding effects that are spurious because noise is being modeled as structure).

I recommend a substantial, deliberate effort to get the *a priori* thinking and models in place and try to obtain more confirmatory results; *then* explore the *post hoc* issues that often arise after seeing the more confirmatory results. Data dredging activities form a continuum, ranging from fairly trivial (venial) to the grievous (mortal). There is often a fine line between dredging and not dredging; my advice is to stay well toward the *a priori* end of the continuum and thus achieve a more confirmatory result. One can always do *post hoc* analyses after the *a priori* analysis; but one can never go from *post hoc* to *a priori*. Why not keep one's options open in this regard?

Grievous data dredging is endemic in the applied literature and still frequently taught or implied in statistics courses without the needed caveats concerning the attendant inferential problems. Rampant rummaging through the data looking for patterns and then "testing" them would be called, in any other human endeavor, *cheating*.

Running all possible models is a thoughtless approach and runs the high risk of finding effects that are, in fact, spurious if only a single model is chosen for inference. If prediction is the objective, model averaging is useful and estimates of precision should include model selection uncertainty; these are subjects to be addressed in later sections of this book. Even in this case, surely one can often rule out many models on *a priori* grounds (e.g., the cement hardening data). There are recent papers in major journals that provide the results of analyses where well over a million models have been run with sample sizes <100. I suspect nearly every result was actually spurious in such cases. Running all possible models is usually a signal of an unthinking science approach.

## 2.6    The Effect of a Flood on European Dippers: Modeling Contrasts

Lebreton et al. (1992) provided a small set of capture–recapture data on the European Dipper (*Cinclus cinclus*). This is a small bird that spends its life along small streams; the data come from eastern France and were collected by Marzolin. The study took place over seven years; thus there are six survival intervals. A flood took place toward the end of the second survival interval and continued into the beginning of the third survival interval. The simple science question asked if survival probability was lower in the two flood years. Note that causation (the flood caused lowered survival) cannot be addressed here as this is an observational study, not a strict experiment.

Some notation is needed; let $\varphi$ be the time-averaged annual survival probability while $\varphi_f$ and $\varphi_{nf}$ be the time-averaged annual survival probabilities for flood and nonflood years, respectively. Specifically, $\varphi$ is the conditional probability that a dipper survives the annual interval and stays on the study area, given it is alive at the beginning of the interval. Finally, we denote the time-averaged probability of capture or recapture as $p$.

### 2.6.1    Traditional Null Hypothesis Testing

Standard practice would be to define a null and alternative hypothesis and their corresponding models. The null hypothesis ($H_0$) would be that there is no effect (exactly no effect) of the flood on annual survival probability, while the alternative hypothesis ($H_a$) would be that the flood did have an effect on annual survival probability. So, we have two models representing the null and alternative hypothesis, respectively:

$H_0$:   $\{\varphi, p\}$ with two unknown parameters
$H_a$:   $\{\varphi_f, \varphi_{nf}, p\}$ with three unknown parameters

The null model is nested within the more general alternative model and this fact allows standard "tests" to be computed to address the issue of a flood effect on annual survival probabilities. This test is done by testing (only) the null hypothesis; the alternative is *not* the subject of the test. If the null is rejected, then, *by default*, the alternative is said to be supported. The alternative hypothesis (the one the investigator usually believes) is never tested.

## 2.6.2  Information-Theoretic Approach

The information-theoretic approach would begin with the same two hypotheses, $\{\varphi, p\}$ and $\{\varphi_f, \varphi_{nf}, p\}$, claiming that these models are only simple approximations to the complex reality. There is no need that the models are nested (they happen to be in this case). The information-theoretic approach asks for measures of relative support (i.e., from the data, empirical) for the two hypotheses. It is not alleged that hypothesis $\{\varphi, p\}$ is exactly true; rather it is a hypothesis and a model that are approximations.

In this early example, perhaps relatively little thought went into the hypotheses to be included in the set – two hypotheses seem "obvious." A little more thought suggests that other hypotheses could be examined (as Bacon and Chamberlin would have wanted):

- Was there a survival effect just the first year of the flood $\{\varphi_{f1}, \varphi_{nf}, p\}$?
- Or just the second year of the flood $\{\varphi_{f2}, \varphi_{nf}, p\}$?
- Or was the recapture probability ($p$) also effected by the flood $\{\varphi_f, \varphi_{nf}, p_f, p_{nf}\}$?
- Or even $\{\varphi, p_f, p_{nf}\}$, where survival was not impacted, but the recapture probabilities were?

Note that few of the models above are nested; thus each model must be tested against the null and this raises the multiple testing problem, a scourge of null hypothesis testing. Traditional tests do not allow much evidence about the relative merit of the four hypotheses/models above.

Thinking hard about hypotheses to be evaluated before data analysis nearly always has its clear rewards. In this simple example, the addition of four more hypotheses was not particularly "heavy mental lifting" but in more challenging problems considerable thought is usually required. We must all do more to encourage a culture of hard thinking and rigor in scientific work. A premium must be placed on thinking, innovation, and creativity – do not expect the computer to tell us what is "important."

Simple problems such as the dipper problem can be effectively addressed with the methods developed in this text; just because the problem is simple does not mean one must use null hypothesis testing methods.


## 2.7   Remarks

Romesburg (2002) wrote a fascinating book about thinking and the creative spirit; I have found this very useful and recommend it.

Chatfield (1995a,b) provides very good guidelines concerning statistical practice. Gotelli and Ellison (2004) provide sage advice on data handling and archiving. Manly's (1992) book covers both sampling and design issues and is easy reading.

Chamberlin's paper is well worth reading after more than 100 years. How many papers in *Science* have been reprinted in the same journal (as was Chamberlain's in 1890 and in 1965)?

Fisher first published on his likelihood approach when he was a third year undergraduate (1912) and a very much extended account in 1922. Likelihood is among the great achievements in statistics (like aspirin in medicine); it is the backbone of statistical thinking, including Bayesian approaches. It might be noted that the Fisher information matrix addresses *precision* (a measure of repeatability) when translated into the covariance matrix, rather than strictly information. Of course, precision is tied to "information." The first book (Edwards 1976) on likelihood was written well after Fisher's fundamental paper on the subject in 1922; this was followed by an expanded treatment in 1992. It is fitting that Edwards was Fisher's last Ph. D. student. Oddly, there are still relatively few books on the subject (good examples include Azzalini 1996; Royall 1997; Severini 2000; Pawitan 2001); like the ubiquitous "delta method"–everyone is supposed to (somehow) know it!

Draper and Smith (1981) provide a review of results found by others that have analyzed the cement hardening data (also see Hald 1952 and Hand 1994). Hendry et al. (2006) give an analysis of the actual data on beak size in Darwin's finches; the hypothetical example here takes their work in a conceptually different modeling direction. Additional results on bovine tuberculosis in feral ferrets in New Zealand are provided by Caley and Hone (2005).

Many papers exist on modeling but there is a clear need for a nice book synthesizing the literature and providing effective examples. Levins (1966), Leamer (1978), Gilchrist (1984), Lehman (1990), Starfield et al. (1990, 1991), Cox (1990, 1995), O'Connor and Spotila (1992), Scheiner and Gurevitch (1993), and Lunneborg (1994). Chatfield (1995a,b, 1991), Nichols (2001), and Shenk and Franklin (2001), and Zuur (2007) offer good introductions into the statistical modeling literature. White and Lubow (2002) provide examples of modeling data from differing sources.

Much of statistical theory is based on an assumption about so-called independence and this is often compromised with data in the life sciences. What is required, in general, is a correct likelihood for the data that reflects any dependence. There is a simple way to handle some lack of independence in making inferences (Sect. 6.1). An easy reading paper on spurious effects and how to minimize these is Anderson et al. (2001a). Inferential problems when using convenience sampling are outlined by Anderson (2001), but see also Hairston (1989) and Eberhardt and Thomas (1991).

## 2.8 Exercises

1. Reread the paper by Caley and Hone (2002).

   a. They demonstrated that estimating the force of infection ($\lambda$) from age-prevalence data is possible and assists in discriminating between alternative hypotheses about routes of disease transmission. Discuss this finding and compare it with similar studies of disease transmission in humans.
   b. Their hypothesis concerning dietary-related transmission from the age of weaning had the best empirical support. Think hard about this and ask if there are logical next steps in understanding the transmission issue. For example, since there was a debate or controversy over this whole issue, what might you want as an opponent?
   c. What would your value judgment be concerning inductive inferences from their sample data to the five populations of ferrets in New Zealand?

2. For those readers with an advanced understanding of mathematical statistics, what worries might you have about getting MLEs of the parameters in the seven models of beak lengths in Darwin's finches? What might be done to avoid problems here?

3. What is "wrong" in merely presenting the eight estimates of flour beetle mortality probability as a function of dose level, either in a table or a simple graph? Why go through the modeling and reparameterization (e.g., the substitution of a submodel involving the parameters $\beta_0$ and $\beta_1$ for $\pi$)? What is the principle here and what are the advantages? (Advanced question).

4. Can you think of any model in the life sciences that is strictly true? What about the physical sciences? Or medicine? Or economics? If possible, ask people in those disciplines for examples of exactly true models in their field. How would a person know with certainty that the true model was in the set, but not know which one is was? Lastly, how do we *know* a model is exactly true? Can you imagine methods that would allow one to determine (e.g., test for) the exact truth of a model? [Probably not a good Ph.D.

project.] This would be a case where the form of the true model would be known, but not its parameters. It seems a shame that true models do not come with their true parameters, making estimation unneeded!

5. Few editors, associate editors, and reviewers seem to be aware of the inferential issues with unadulterated data dredging. They seem to believe that all analysis results are created equal and it makes no difference if the hypothesis was posed before or after data analysis. Discuss this issue. How can this issue be improved so our science moves ahead more rapidly?

6. Linhart and Zucchini (1986) analyzed data on weekly storm events at a botanical garden in Durban, South Africa. They had data over 47 consecutive years and were interested in prediction of weekly storm events (i.e., $i = 1$, $2, \ldots, 52$ weeks). They knew that an estimator of the probability of a storm in week $i$ was $\hat{p}_i = y_i / 47$, where $y_i$ is the number of storms in week $i$. Thus, they computed the binomial estimator (an MLE) for all 52 weeks. Critique this approach. What is "wrong" here?

# 3

# Information Theory and Entropy

*Solomon Kullback* (1907–1994) was born in Brooklyn, New York, USA, and graduated from the City College of New York in 1927, received an M.A. degree in mathematics in 1929, and completed a Ph.D. in mathematics from the George Washington University in 1934. Kully as he was known to all who knew him, had two major careers: one in the Defense Department (1930–1962) and the other in the Department of Statistics at George Washington University (1962–1972). He was chairman of the Statistics Department from 1964–1972. Much of his professional life was spent in the National Security Agency and most of his work during this time is still classified. Most of his studies on information theory were done during this time. Many of his results up to 1958 were published in his 1959 book, *"Information Theory and Statistics."* Additional details on Kullback may be found in Greenhouse (1994) and Anonymous (1997).

When we receive something that decreases our uncertainty about the state of the world, it is called *information*. Information is like "news," it informs. Information is not directly related to physical quantities. Information is not material and is not a form of energy, but it can be stored and communicated using material or energy means. It cannot be measured with instruments but can be defined in terms of a probability distribution. Information is a decrease in uncertainty.

This textbook is about a relatively new approach to empirical science called "information-theoretic." The name comes from the fact that the foundation originates in "information theory"; a set of fundamental discoveries made largely during World War II with many important extensions since that time. One exciting discovery is the ability to actually quantify *information* and this has led to countless breakthroughs that affect many things in our daily lives (e.g., cell phone and global positioning system technologies). One might think of information theory as being things like coding and encrypting theory and signal transmission, but it is far more general than these subjects.

Allowing "data analysis" to hook up with information theory has had substantial advantages and statistical scientists are still trying to exploit this combination. The concepts and practical use of the information-theoretic approach are simpler than that of hypothesis testing, and much easier than Bayesian approaches to data analysis.

**Before proceeding further, I want to summarize the necessary "setting." This setting will set the tone for all of the following material.** I will assume the investigator has a carefully considered science question and has proposed R hypotheses (the "multiple working hypotheses"), all of which are deemed plausible. A mathematical model (probability distribution) has been derived to well represent each of the $R$ science hypotheses. Estimates of model parameters ($\theta$) and their variance–covariance matrix ($\Sigma$) have been made under either a least squares (LS) or maximum likelihood (ML) framework. In either case, other relevant statistics have also been computed (adjusted $R^2$, residual analyses, goodness-of-fit tests, etc.). Then, under the LS framework, one has the residual sum of squares (RSS), while under a likelihood framework, one has the value of the log-likelihood function at its maximum point. **This value (either RSS or max log($\mathcal{L}$)) is our starting point and allows answers to some of the relevant questions of interest to the investigator, such as:**

- Given the data, which science hypothesis has the most empirical support (and by how much)?
- What is the ranking of the $R$ hypotheses, given the data?
- What is the probability of, say, hypothesis 4, given the data and the set of hypotheses?
- What is the (relative) likelihood, say, of hypothesis 2 vs. hypothesis 5?
- How can rigorous inference be made from all the hypotheses (and their models) in the candidate set? This is multimodel inference.

## 3.1   Kullback–Leibler Information

The scope of theory and methods that might be classed as "information theory" is very large. I will focus primarily on Kullback–Leibler information and this comes from a famous paper by Soloman Kullback and Richard Leibler published in 1951. Their work was done during WWII and published soon after the termination of the war.

---

### Kullback–Leibler Information

In the context of this book, Kullback–Leibler (K–L) information is a function denoted as "*I*" for information. This function has two arguments: *f* represents full reality or "truth" and *g* is a model. Then, **K–L information *I(f, g)* is the**

**"information" lost when the model *g* is used to approximate full reality, *f*.**

An equivalent, and very useful, interpretation of *I(f, g)* is the

**"distance" from the approximating model *g* to full reality, *f*.**

Under either interpretation, we seek to find a candidate model that minimizes *I(f, g)*, over the hypothesis set, represented by models.

---

Thus, if one had a set of five hypotheses, each represented by a model, *I(f, g)* would be computed for each of the five. The model with the smallest information loss would be the best model and, therefore, would represent the best hypothesis. The model *g* has its parameters given; there is no estimation and no data involved at this point (this will change as we go forward).

Alternatively, one could interpret the model with the smallest *I(f, g)* value as being "closest" to full reality. Thus, when a "best model" is mentioned, the "best" will stem from the concept of the smallest information loss or a model being closest to full reality. This is a conceptually simple, yet powerful, approach. The idea of a "distance" between a model and full reality seems compelling.

Kullback–Leibler information is defined by the unpleasant-looking integral for continuous distributions (e.g., the normal or gamma):

$$I(f,g) = \int f(x) \log\left(\frac{f(x)}{g(x|\theta)}\right) dx.$$

K–L information is defined as the summation for discrete distributions (e.g., Poisson, binomial, or multinomial):

$$I(f,g) = \sum_{i=1}^{k} p_i \log\left(\frac{p_i}{\pi_i}\right).$$

Here, there are $k$ possible outcomes of the underlying random variable; the true probability of the $i$th outcome is given by $p_i$, while the $\pi_1,\ldots,\pi_k$ constitute the approximating probability distribution (i.e., the approximating model). In the discrete case, we have $0 < p_i < 1$, $0 < \pi_i < 1$, and $\sum p_i = \sum \pi_i = 1$. Hence, here $f$ and $g$ correspond to the $p_i$ and $\pi_i$, respectively. In the following material, we will generally think of K–L information in the continuous case and use the notation $f$ and $g$ for simplicity.

Some readers might start to "lose it" thinking that they must compute K–L information loss for each model in the set. It turns out that *I(f, g)* cannot be used directly because it requires knowledge of full reality ($f$) and the parameters ($\theta$) in the approximating models, $g_i$; we will never have knowledge of these entities in real problems. We will see that K–L information can be easily *estimated*, without advanced mathematics (although the derivation is very deeply mathematical). This estimation requires data relevant to the science question.

Kullback–Leibler information is the most fundamental of all information measures in the sense of being derived from minimal assumptions and its additivity property. It can be viewed as a quantitative measure of the *inefficiency* of assuming a model $g$ when truth is $f$. Again, one wants to select a model from the set that minimizes inefficiency. While the Kullback–Leibler distance can be conceptualized as a "distance" between $f$ and $g$, strictly speaking this is a measure of "discrepancy." It is not a simple distance because the measure from $f$ to $g$ is not the same as the measure from $g$ to $f$ – it is a "directed" or "oriented" distance.

## 3.2    Linking Information Theory to Statistical Theory

We usually think that "data analysis" is tied in with the subject of "statistics." How are statistical principles linked with information theory, and K–L information in particular? This linkage was the genius of Hirotugu Akaike in an incredible discovery first published in 1973.

A glimpse into the linkage between information and entropy and their relationship to mathematical statistics is given below; a full and technical derivation appears in Burnham and Anderson (2002:Chap. 7). I urge people to wade through this to gain a notion of the derivation. In particular, when there are unknown parameters to be estimated from data, the criterion must change. This change is introduced in the derivation to follow:

Akaike's main steps started by using a property of logarithms (i.e., $\log(A/B) = \log(A) - \log(B)$) to rewrite K–L information as

$$I(f,g) = \int f(x)\log(f(x))dx - \int f(x)\log(g(x\,|\,\theta))dx.$$

Both terms on the right-hand side are statistical expectations (Appendix B) with respect to $f$ (truth). Thus, K–L information can be expressed as

$$I(f,g) = E_f[\log(f(x))] - E_f[\log(g(x\,|\,\theta))],$$

each expectation with respect to the true distribution $f$. This last expression provides insights into the derivation of AIC. Note that no approximations have been made, no parameters have been estimated and there are no data at this point; K–L information has merely been re-expressed.

The first expectation is a constant that depends only on the conceptual true distribution and it is not clearly known. However, this term is constant across the model set. In other words, the expectation of $[\log(f(x))]$ does not change from model to model; it is a constant. Thus, we are left with only the second expectation,

$$I(f,g) - C = -E_f[\log(g(x \mid \theta))].$$

The constant term ($C$) can be made to vanish in a subsequent step (Chap. 4). The question now is if we can somehow compute or estimate $E_f[\log(g(x \mid \theta))]$. The short answer is no as the criterion or target must be altered to achieve a useful result and this will require data.

Kullback–Leibler information or distance $I(f, g)$ is on a true ratio scale, where there is a true zero. In contrast, $-E_f[\log(g(x \mid \theta))] = -\int f(x)\log(g(x \mid \theta))dx$ is on an interval scale and lacks a true zero, because of the constant (above). A difference of magnitude $D$ means the same thing anywhere on the scale. Thus, $D = 10 = 12 - 2 = 1012 - 1002$; a difference of 10 means the same thing anywhere on the interval scale. Then, $10 = V_1 - V_2$, regardless of the size of $V_1$ and $V_2$. A large sample size magnifies the separation of research hypotheses and the models used to represent them. Adequate sample size conveys a wide variety of advantages in making valid inferences (e.g., improved estimates of $E_f[\log(g(x \mid \theta))]$).

## 3.3    Akaike's Information Criterion

Akaike introduced his information-theoretic approach in a series of papers in the mid-1970s as a theoretical basis for model selection. He followed this pivotal discovery with several related contributions beginning in the early 1980s and classified these as falling under the *entropy maximization principle*. This world class discovery opened the door for the development of relatively simple methods for applied problems, ranging from simple to quite complex, but based on very deep theories – entropy and K–L information theory on the one hand and Fisher's likelihood theory (see Appendix A) on the other.

Akaike's (1973) seminal paper used Kullback–Leibler information as a fundamental basis for model selection and recognized model parameters must be estimated from data and there is substantial uncertainty in this estimation. The estimation of parameters represents a major distinction from the case where model parameters are assumed to be known. Akaike's finding of a relation between the K–L information and the maximized log-likelihood has allowed major practical and theoretical advances in model selection and the analysis of complex data sets. deLeeuw (1992) said it well, "Akaike found a formal relationship between Boltzmann's entropy and Kullback–Leibler information (dominant paradigms in information and coding theory) and maximum likelihood (the dominant paradigm is statistics)."

Akaike's next step was stymied as no way could be found to compute or estimate the second term, $E_f[\log(g(x|\theta))]$. However, the expectation of this quantity led to a major breakthrough. Data enter the derivation and allow parameter estimates ($\hat{\theta}$)Akaike found that he could not estimate K–L, but he could estimate the expectation of K–L information. This second expectation is over the data (denote these data as $y$)

$$E_f[\log(g(x|\hat{\theta}))],$$

where the estimates $\hat{\theta}$ are based on the data ($y$).

---

**The Modified Target**

Akaike showed that the critical issue became the estimation of

$$E_y E_x[\log(g(x|\hat{\theta}(y)))].$$

This double expectation, both with respect to truth $f$, is the target of all model selection approaches based on K–L information. This notation makes it clear that the first (outer) expectation is over the data ($y$) and these data allow estimates of the unknown model parameters. Thus, we now have modified the target of relevance here due to the need for data to estimate model parameters. The proper criterion for model selection relates to the *fitted* model. The modification required is *expected* K–L information; Akaike called this a "predictive likelihood."

Akaike realized that this complex entity was closely related to the log-likelihood function at its maximum. However, the maximized log-likelihood is biased upward as an estimator of this quantity. Akaike found that, under certain conditions, this bias is approximately equal to $K$, the number of estimable parameters in the approximating model. This is an asymptotic (meaning as sample size increases to infinity) result of fundamental importance.

Thus under mild conditions, an asymptotically unbiased estimator of

$$E_y E_x[\log(g(x|\hat{\theta}(y)))] = \log(\mathcal{L}(\hat{\theta}|\text{data}) - K.$$

---

This stunning result links expected K–L information to the maximized log-likelihood ($\log(\mathcal{L})$) corrected for bias. The important linkage is summarized as,

**negentropy = K–L information and E(K–L information) = log($\mathcal{L}$) – K**

*thermodynamics*          *information theory*                              *statistics*

Akaike's final step defined "*an information criterion*" (AIC) by multiplying both terms through by $-2$ ("taking historical reasons into account"). Thus, *both* terms in $\log(\mathcal{L}(\hat{\theta}|data)) - K$ were multiplied by $-2$ to get

$$\text{AIC} = -2\log(\mathcal{L}(\hat{\theta})|\text{data}) + 2K.$$

This has become known as *Akaike's Information Criterion* or AIC. AIC has a strong theoretical underpinning, based on entropy and expected Kullback–Leibler information. Akaike's inferential breakthrough was finding that the maximized log-likelihood could be used to estimate the expected (averaged) K–L distance between the approximating model and the true generating mechanism. The expectation of the logarithm of $f(x)$ drops out as a constant across models, independent of the data.

In practice, one computes AIC for each of the models in the set and then selects the model that yields the smallest value of AIC for inference. One justifies this selection because that selected model minimizes the information lost when approximating full reality by a fitted (i.e., parameters estimated from the data using, for example, ML or LS methods) model. Said another way, that selected model is "closest" to full reality, given the data. This approach seems a very natural, simple concept; select the approximating model that is closest to the unknown reality.

It might be argued that I should have merely defined $l = \log(\mathcal{L}(\theta|\text{data,model}))$; then AIC $= -2l + 2K$, making the criterion appear more simple. While this may have advantages, I believe the full notation works for the reader and helps in understanding exactly what is meant. The full notation, or abbreviations such as $\log(\mathcal{L}(\theta|x,g_i))$, makes it implicit that the log-likelihood is a function of (only) the parameters ($\theta$); while the data ($x$) and model ($g_i$, say multinomial) must be *given* (i.e., known). These distinctions become more important when we introduce the concept of a likelihood of a model, given the data: $\mathcal{L}(g_i|\text{data})$ in Chap. 4. Both concepts are fundamental and useful in a host of ways in this book and the notation serves an important purpose here.

### 3.3.1  The Bias Correction Term

Correction of estimators for bias has a long history in statistics. The usual estimator of the variance is a ready example

$$\text{variance} = \frac{\sum (x_i - \hat{\mu})^2}{n-1},$$

where the subtraction of 1 from the sample size ($n$) in the denominator corrects for a small sample bias (note that as $n$ gets large the bias correction becomes unimportant). The bias correction term ($K$ = the number of estimable parameters), above, is a special case of a more general result derived by Takeuchi (1976) and described in Sect. 3.9.1. AIC is a special case of Takeuchi's Information Criterion (TIC) and is, itself, a parsimonious approach to the estimation of expected K–L information.

### 3.3.2  Why Multiply by −2?

Akaike multiplied the bias-corrected log-likelihood by −2 for "historical reasons." It is a well-known statistical result that −2 times the logarithm of the ratio of two maximized likelihood values is asymptotically chi-square distributed under certain conditions and assumptions (this is the likelihood ratio test). The term −2 occurs in other statistical contexts, and so it was not unreasonable that Akaike performed this simple operation to get his AIC. Three points frequently arise and I will note these here.

First, the model associated with the minimum AIC remains unchanged had the bias-corrected log-likelihood (i.e., $\log(\mathcal{L}) - K$) been multiplied by −0.17, −51.3, −3.14159, or any other negative number. Thus, the minimization is not changed by the multiplication of *both* terms by any negative constant; Akaike merely chose −2. Second, some investigators have not realized the formal link between expected K–L information and AIC and believed, then, that the number 2 in (*only*) the second term in AIC was somehow arbitrary and that other multipliers should also be considered. This error has led to considerable confusion in the technical literature; −K is the asymptotic bias correction and is not arbitrary. Akaike chose to work with $-2\log(\mathcal{L})$, rather than $\log(\mathcal{L})$; thus the term $+ 2K$ is theoretically correct for large sample size. As long as *both* terms (the log-likelihood and the bias correction term) are multiplied by the same negative constant, the model where the criterion is minimized is unchanged and there is nothing arbitrary. Third, $-2\log(\mathcal{L})$ is termed "deviance" in mathematical statistics. People with a statistical background immediately interpret deviance as a way to quantify lack of fit and they then view AIC as simply "deviance + 2K." I suspect that this was Akaike's thinking when he multiplied through by −2; that is simply, "deviance penalized by 2K to correct for asymptotic bias."

### 3.3.3   Parsimony is Achieved as a by-Product

AIC is linked directly to the estimation of expected K–L information. The derivation itself was not based on the concept of parsimony. It was after Akaike's elegant derivation of AIC that people noticed a heuristic interpretation that was interesting and allowed insight into how parsimony is enforced with AIC. The best model is closest to full reality and, therefore, the goal is to find the model where AIC is smallest. The first term (the deviance) in AIC

$$\text{AIC} = -2\log(\mathcal{L}(\hat{\theta})|\, x) + 2K$$

is a measure of lack of model fit, and can be made smaller by adding more parameters in the model $g_i$. Thus, for a fixed data set, the further addition of parameters in a model $g_i$ will allow it to fit better. However, when these added parameters must be estimated (rather than known or "given"), further uncertainty is added to the *estimation* of expected K–L information or distance. At some point, the addition of still more estimated parameters will have the opposite effect and the estimate of expected K–L information will increase because "noise" is then being modeled as if it were structural. The second term in AIC (2K) then functions as a "penalty" for adding more parameters in the model.

Thus, the penalty terms (2K) gets larger as more parameters are added. One can see that there is a tension between the deviance and the penalty term as the number of parameters is increased – a trade-off.

Without a proper penalty term the best model would nearly always be the largest model in the set, because adding more and more parameters to be estimated from the fixed amount of data would be without "cost" (i.e., no penalty). The result would be models that are overfit, have low precision, and risk spurious effects because noise is being modeled as structure.

This heuristic explanation does not do justice to the much deeper theoretical basis for AIC (i.e., the link with expected K–L information). However, the advantage of adding parameters and the concomitant disadvantage of adding still more parameters suggests a trade-off. This is the trade-off between bias and variance or the trade-off between underfitting and overfitting that is the Principle of Parsimony (see Sect. 2.4). Note that parsimony was not a condition leading to AIC, instead parsimony appears almost as a by-product of the end result of the derivation of AIC from expected K–L information.

Inferences for a given data set are conditional on sample size. We must admit that if much more data were available, then further effects could probably be found and supported. "Truth" is elusive; model selection tells us what inferences the data support, not what full reality might be. Full reality cannot be found using a finite data set.

### 3.3.4   Simple vs. Complex Models

Data analysis involves the critical question, "how complex a model will the data support?" and the proper trade-off between underfitting and overfitting. This dilemma has had a long history in the analysis of data and model based inference. As biologists, we think certain variables and structure must be in a 'good model' often without recognition that putting in too many variables and too much structure introduces large uncertainties, particularly when sample size is relatively small or even moderate. In addition, interpretability is often decreased as the number of parameters increases.

As biologists, we have a strong tendency to want to build models of the information in the data that are too complex (overfit). This is a parsimony issue that is central to proper model selection. One cannot rely on intuition to judge a proper trade-off between under- and overfitting, a criterion based on deep theory is needed. Expected K–L information and AIC provide the basis for a rigorous trade-off. This seems a very natural, simple concept; *select the fitted approximating model that is estimated, on average, to be closest to the unknown full reality, f.*

Ideal model selection results in not just a good fitting model, but a model with good out-of-sample prediction performance. This is a tall order. The selected model should have good achieved confidence interval coverage for the estimators in the model and small predictive mean squared errors (PMSE).

### 3.3.5  AIC Scale

As defined, AIC is strictly positive. However, during an analysis, it is common to omit mathematical terms that are constant across models and such shortcuts can result in negative values of AIC. Computing AIC from regression statistics often results in negative AIC values. This creates no problem, one just identifies the model with the smallest value of AIC and declares it is the model estimated to be the best. This fitted model is estimated to be "closest" to full reality and is a good approximation for the information in the data, relative to the other models considered. For example,

| Model | AIC |
|-------|-----|
| $g_1$ | 1,400 |
| $g_2$ | 1,570 |
| $g_3$ | 1,390 |
| $g_4$ | 1,415. |

One would select model $g_3$ as the basis for inference as it has the smallest AIC value; meaning that it is estimated to be closest to full reality. Because these values are on a relative scale, one could subtract, say, 2,000 from each and have the following rescaled AIC values: −600, −430, −610, and −585. The rank of each model is not changed by the rescaling; the ranks, in each case remain $g_3$ (best), $g_1$, $g_4$, and $g_2$ (worst). I have seen AIC values that range from −80,000 to as high as 340,000 in different scientific applications. It is not the absolute size of the AIC value, it is the *relative* values, and particularly the differences, that are important (Chap. 4).

## 3.4  A Second-Order Bias Correction: AICc

---

**Second-Order Bias Correction: AICc**

Akaike derived an asymptotically unbiased estimator of expected K–L information; however, AIC may perform poorly if there are too many estimated parameters in relation to the size of the sample. A second-order variant of AIC has been developed and **it is important to use this criterion in practice:**

$$\text{AICc} = -2\log(\mathcal{L}(\hat{\theta})) + 2K\left(\frac{n}{n-K-1}\right),$$

where $n$ is sample size. This can be rewritten as

$$\text{AICc} = -2\log(\mathcal{L}(\hat{\theta})) + 2K + \frac{2K(K+1)}{n-K-1}$$

or equivalently

$$\text{AICc} = \text{AIC} + \frac{2K(K+1)}{n-K-1}.$$

---

AICc merely has an additional bias correction term. If $n$ is large (asymptotic) with respect to $K$, then the second-order correction is negligible and AICc converges to AIC. AICc was derived under Gaussian assumptions and is weakly dependent on this assumption. Other model-specific assumptions can be made and this might be worthwhile in data analysis where there are severe controversies or consequences (Burnham and Anderson 2002: Chap. 7). *The use of AICc is highly recommended in practice; do not use just AIC*.

## 3.5   Regression Analysis

Least squares regression is a very useful approach to modeling. Here, model selection is often thought of as "variable selection." It is easy to move from regression statistics such as the residual sum of squares (RSS) to the log-likelihood function at its maximum point; this allows one to use AICc. Note, LS and ML provide exactly the same estimates of the $\beta_i$ in linear models; however, the estimates of the residual variance $\sigma^2$ can differ appreciably if sample size is small.

---

**Mapping the RSS into the Maximized Log-Likelihood**

The material to this point has been based on likelihood theory (Appendix A) as it is a very general approach. In the special case of LS estimation ("regression") with normally distributed errors, and apart from a constant, we have

$$\log(\mathcal{L}) = -\frac{n}{2}\cdot\log(\hat{\sigma}^2).$$

Substituting this expression, AICc for use in LS models can be expressed as

$$\text{AICc} = n\log(\hat{\sigma}^2) + 2K\left(\frac{n}{n-K-1}\right),$$

where $\hat{\sigma}^2 = \sum \hat{\varepsilon}_i^2 / n$ (the MLE) and $\hat{\varepsilon}_i$ are the estimated residuals for a particular candidate model.

A common (but minor) mistake is to take the LS estimate of $\sigma^2$ from the computer output, instead of the ML estimate (above). In regression models, $K$ is the total number of estimated parameters, including the intercept and $\sigma^2$. The value of $K$ is sometimes computed incorrectly as either $\beta_0$ or $\sigma^2$ are mistakenly ignored in obtaining $K$. AICc is easy to compute from the results of LS estimation in the case of linear models. It is not uncommon to see computer software that computes simple AIC value incorrectly; few packages provide AICc; however, this can be computed easily manually.

---

Given a set of candidate models $g_i$, with parameters to be estimated from the observed data, the model which minimizes the predictive expectation is "closest" to full reality ($f$) and is to be preferred as a basis for inference. AICc allows an estimate as to which model is best for a given data set; however, a different best model might be selected if another (replicate) data set was available. These are stochastic biological processes, often with relatively high levels of complexity, we must admit to uncertainty and try to quantify it. This condition is called "model selection uncertainty." We must also admit that if much more data were available, then further effects could probably be found and supported. "Truth" is elusive; proper model selection helps us understand what inferences the data support.

AICc attempts to reformulate the problem explicitly as a problem of *approximation* of the true structure (probably infinite dimensional) by a *model*. Model selection then becomes simply finding the model where AICc is minimized. I will show in a later chapter that model selection is much more than this.

AICc selection is objective and represents a very different paradigm to that of null hypothesis testing and is free from the arbitrary $\alpha$ levels, the multiple testing problem, and the fact that many candidate models are not nested. The problem of what model to use is inherently not a null hypothesis testing problem.

The fact that AIC allows a simple comparison of models does not justify the comparison of all possible models. If one had 10 variables, then there are 1,024 possible models, even if interactions and squared or cubed terms are excluded. If sample size is $n \leq 1,000$, overfitting is almost a certainty. It is simply not sensible to consider such a large number of models because an overfit model will almost surely result and the science of the problem has been lost. Even in a very exploratory analysis it seems like poor practice to consider all possible models; surely some science can be brought to bear on such an unthinking approach. I continue to see papers published where tens of thousands or even millions of models are fit and evaluated; this represents a foolish approach and virtually guarantees spurious effects and absurdities.

As a generally useful rule, when the number of models ($R$) exceeds the sample size ($n$), one is asking for serious inferential difficulties. I advise people to think first about their set of *a priori* science hypotheses; these will typically be relatively few in number. A focus on models is the result of computer software that is very powerful, but unthinking.

## 3.6    Additional Important Points

### 3.6.1    Differences Among AICc Values

Often data do not support only one model as clearly best for data analysis (i.e., little or no model selection uncertainty). Instead, suppose three models are essentially tied for best, while another subset of models is clearly not appropriate (either under- or overfit). Such virtual "ties" for the estimated best model must be carefully considered and admitted. The inability to ferret out a single best model is not a defect of AICc or any other selection criterion, rather, it is an indication that the data are simply inadequate to reach such a strong inference.

It is perfectly reasonable that several models would serve nearly equally well in approximating the information in a set of data. Inference must admit that there are sometimes competing hypotheses and the data do not support selecting only one. Large sample sizes often reduce close ties among models in the set. The issue of competing models is especially relevant in including model selection uncertainty into estimators of precision and model averaging (Chap. 5).

Consider studies of *Plasmodium* infection of children in tropical Africa and data from two different sites have been modeled and fitted. The best model for the eastern site has AIC = 104, whereas the best model for the western site has AIC = 231. Are the models better for the western site? Perhaps, however, just the fact that the best model for the western site has a larger AIC value is *not* evidence of this. AIC values are functions of sample size and this precludes comparing AIC values across data sets.

### 3.6.2 Nested vs. Nonnested Models

The focus should be on the science hypotheses deemed to be of interest. Modeling of these hypotheses should not be constrained to only models that are nested. AICc can be used for nonnested models and this is an important feature because likelihood ratio tests are valid only for nested models. The ranking of models using AICc helps clarify the importance of modeling.

### 3.6.3 Data and Response Variable Must Remain Fixed

It is important that the data are fixed prior to data analysis. One cannot switch from a full data set to one where some "outliers" have been omitted in the middle of the analysis. It would be senseless to evaluate two hypotheses using data $x$ and the remaining four hypotheses using a somewhat different data set. The fixed nature of the data is implied in the shorthand notation for models: $g(\theta \,|\text{data})$, the model as a function of the known parameters $(\theta)$, given the (fixed) data $(x)$.

Some analyses can be done on either the raw data or some grouping of the raw data (e.g., histogram classes). In such cases, one must be consistent in performing the analysis on one data type or the other, not a mixture of both types. Any grouping of the raw data loses some information, thus grouping should be carefully considered.

If $Y$ is the response variable of interest, it must also be kept fixed during the analysis. One cannot evaluate models of $Y$ and then switch to models of $\log(Y)$ or $\sqrt{Y}$. Having a mix of response variables in the model set is an "apples and oranges" issue. Such changes make the AICc values uninterpretable; more importantly, the science problem is muddied. For example, presence–absence data on some plant species cannot be compared to counts of that plant on a series of plots.

### 3.6.4    AICc is not a "Test"

Information-theoretic approaches do not constitute a statistical "test" of any sort (see Appendix C). There are no test statistics, assumed asymptotic sampling distributions, arbitrary $\alpha$-levels, $P$-values, and arbitrary decision about "statistical significance." Instead, there are numerical values that represent the scientific evidence (Chaps. 4 and 5), often followed by value judgments made by the investigators and perhaps others.

It is poor practice to mix evidence from information-theoretic approaches with the results of null hypothesis testing, even though this is a common mistake in the published literature. One sees cases where the models are ranked using AICc and then a "test" is carried out to see if the best model is "significantly better" than the second-best model. This is seriously wrong on several different technical levels and I advise against it. Null hypothesis testing and information-theoretic results are like oil and water; they do not mix well.

### 3.6.5    Data Dredging Using AICc

Ideally science hypotheses and their models are available prior to data analysis and, ideally, prior to data collection. These *a priori* considerations led to a confirmatory result. Following that, I often encourage some *post hoc* examination of the data using hypotheses and models suggested by the *a priori* results. Such after-thoughts are often called "data dredging." I do not condone the use of information-theoretic criteria in data dredging, even in the early phases of exploratory analysis. For example, one might start with 8–10 models, compute AICc for each, and note that several of the better models each have a gender effect. Based on these findings, another 4–7 models are derived to include a gender effect. After computing AICc for these models, the analyst notes that several of these models have a trend in time for some parameter set; thus more models with this effect are derived, and so on. This strategy constitutes traditional data dredging but using an information theoretic criteria instead of some form of test statistic or visual inspection of plots of the intermediate results. I recognize that others have a more lenient attitude toward blatant data dredging. I think investigators should understand the negative aspects of data dredging and try to minimize this activity.

### 3.6.6    Keep all the Model Terms

It is good practice to retain all the terms in the log-likelihood in order for AICc to be comparable across models. This is particularly important for nonnested models (e.g., the nine models of Flather, Sect. 3.9.6) and in cases where different error distributions are used (e.g., log-normal vs. gamma). If several computer programs are used to get the MLEs and the maximum $\log(\mathcal{L})$, then one is at risk that some terms in one model were dropped, while these terms were not dropped in other models. This is a rather technical issue: Burnham and Anderson (2002, Sect. 6.7) provide some insights and examples.

### 3.6.7 Missing Data

A subtle point relates to data sets where a few values of the response variable or predictor variables are missing. Such missing values can arise for a host of reasons, including loss, unreadable recording, and deletion of values judged to be incorrect. If a value or two are missing from a large data set, perhaps no harm is done. However, if the missing values are numerous at all then more careful consideration is called for. In particular, if some values for covariates are missing, this can also lead to important issues, including the fact that some software may either stop or give erroneous results. There are *ad hoc* routines for assigning "innocent" values to be used in place of the missing values; these could be considered. There are a variety of Bayesian "imputation" techniques that have merit; these are far beyond the scope of this text. The real moral here is to collect data with utmost care and in doing so, avoid issues with missing data.

### 3.6.8 The "Pretending Variable"

Putting aside the second-order correction for bias for a moment, AIC is just $-2\log(\mathcal{L}) + 2K$ or deviance $+ 2K$. The addition of each new parameter suffers a "penalty" of 2. Now, consider the case where model A has $K$ parameters and model B has $K + 1$ parameters (i.e., one additional parameter). Occasionally, we find that model B is about 2 units from model A and thus, we would view model B as a good model – it *is* a good model. Problems arise when the two AIC values are about 2 units apart but the deviance is little changed by the addition of a variable or parameter in model B. In this case, the additional variable does not contribute to a better fit, instead, it is a "good" model only because the bias correction term is only 2 (i.e., $2 \times 1$). This should not be taken as evidence that the new parameter (and the variable it is associated with) is important. The new parameter is only "pretending" to be important; to confirm this, one can examine the estimate of the parameter (perhaps a regression coefficient $\beta$) and its confidence interval. However, the real clue here is that the deviance did not change and this is an indication that the model fit did not improve.

  I will call this issue a "pretending variable" as a noninformative variable enters as one additional parameter and therefore incurs only a small "penalty" of about 2, but does not increase the log-likelihood (or decrease the deviance). Is this model (B) a good model? YES. Can we take this result to further imply that the added variable is important? NO. Thus, scientists must examine the table of model results to be sure that added variables increase the log-likelihood values. Pretending variables may arise for any models $i$ and $j$ in the set where the difference in AIC values increase by about 2. Less commonly, a model (call it C) will add two parameters and its added penalty is 4 (still a decent model). However, unless there is a change in the log-likelihood, the two new variables or parameters are only "pretending" to be important. Finally, when using AICc, the values are a bit different from 2 or 4; still a focus on the log-likelihood or deviance is advised, to be sure that the fit has improved.

## 3.7   Cement Hardening Data

The computation of AICc from the cement hardening data from Sect. 2.2.1 is shown in the table below:

| Model | $K$ | $\hat{\sigma}^2$ | $\log(\mathcal{L})$ | AICc | Rank |
|---|---|---|---|---|---|
| {mean} | 2 | 208.91 | −34.72 | 74.64 | 5 |
| {12} | 4 | 4.45 | −9.704 | 32.41 | 1 |
| {12 1*2} | 5 | 4.40 | −9.626 | 37.82 | 2 |
| {34} | 4 | 13.53 | −16.927 | 46.85 | 3 |
| {34 3*4} | 5 | 12.42 | −16.376 | 51.32 | 4 |

PROC REG (SAS Institute 1985) was used to compute the residual sum of squares (RSS), the LS estimates of the $\beta$ parameters, and the standard errors of the parameter estimates for each model. The MLE of the residual variance is $\hat{\sigma}^2 = RSS/n$, where the sample size $(n) = 13$. AICc $= n\log(\hat{\sigma}^2) + 2K + 2K(K + 1)/(n - K - 1)$ was used. The calculations can be illustrated using the information from the {mean} model above where $K = 2$. The MLE of the residual variance is 208.91, thus the first term in AICc is 13 log(208.91) = 69.444, the second term is 2·2 = 4, and the third term is (2·2·3)/(13 − 3) = 1.2. Summing the three terms leads to 74.64. The computations are easy but the reader should compute a few more entries in the table above to be sure they understand the notation and procedure. Note, log-likelihood values are usually negative, while AICc values are generally positive.

### 3.7.1   Interpreting AICc Values

AICc is an estimator of expected K–L information and we seek the fitted model where the information loss is minimal. Said another way, we seek the model where the estimated distance to full reality is as small as possible; this is the model with the smallest AICc value. The model that is estimated to be closest to full reality is referred to as the "best model." This best model is {12} from the table above, namely,

$$E(Y) = \beta_0 + \beta_1(x_1) + \beta_2(x_2)$$

with four parameters ($K = 4 = \beta_0, \beta_1, \beta_2,$ and $\sigma^2$).
   The {mean} model in the table is just the mean and variance of the response variable, thus only two parameters are estimated, $\beta_0$ and $\sigma^2$. The model notation {12 1*2} denotes the model

$$E(Y) = \beta_0 + \beta_1(x_1) + \beta_2(x_2) + \beta_3(x_1 * x_2)$$

where an interaction term is introduced. This model is estimated to be the second best model; however, one can quickly see that the interaction is not an important predictor by examining the MLEs and their standard errors:

| Parameter | MLE | $\hat{se}(\hat{\beta})$ |
|---|---|---|
| $\beta_1$ | 1.275 | 0.581 |
| $\beta_2$ | 0.636 | 0.091 |
| $\beta_3$ | **0.004** | **0.012** |

Note that the estimated standard error on $\hat{\beta}_3$ is three times larger than the MLE, thus it certainly seems to be unimportant. This result allows an important, but subtle, point to be made here. Let us ask two questions from the information above. First, is the interaction model {12 1*2} a relatively good model? The answer is YES, this can be seen from the table of AICc values. Second, does this answer imply that the interaction term is an important predictor? The answer is *no*; to judge its importance one needs to examine the standard error of the estimate and a confidence interval (i.e, −0.020 to 0.028 for $\beta_3$). This interval was computed as $\hat{\beta}_3 \pm 2 \times \hat{se}(\hat{\beta}_3)$ and is essentially centered on zero and fails to support the hypothesized importance of $\beta_3$.

Another thing to note is that the scale (i.e., the size, Sect. 3.3.5) of the AICc values is unimportant. One cannot look at the AICc value for the third model (37.82) and judge weather it is "too big" or not "big enough." These AICc values have unknown constants associated with them and are functions of sample size. It is the relative values of AICc that are relevant. In fact, we will see in Chap. 4 that it is the *differences* in AICc values that become the basis for extended inferences.

Of course AICc allows a quick ranking of the five hypotheses, represented by the five models. The ranks are (from estimated best to worst): $g_2$, $g_3$, $g_4$, $g_5$, and $g_1$ for this simple example. Models {34} and {34 3*4} are poor and the mean-only model is very poor in the rankings. The ability to rank science hypotheses is almost always important; however, it will be seen that far more information can be gained using methods introduced in the following chapter.

### 3.7.2   What if all the Models are Bad?

If all five models are essentially worthless, AICc will still rank them; thus, one must have some way to measure the worth of the best model or the global model. In regression, a natural measure of the worth of a model is the adjusted $R^2$ value. In other contexts, one can use a method outlined by Nagelkerke (1991) for a likelihood-based analysis. In this case of cement hardening, the model estimated to be the best in the set was model {12} with an adjusted $R^2 = 0.974$. This suggests that the best model is quite good, at least for these data.

If this best model and its MLEs were used with a new, replicate data set, one would find that the adjusted $R^2$ would be substantially lower than 0.974. Adjusted $R^2$ exaggerates our notions about the out-of-sample predictive ability of models fit to a given data set. The derivation of AICc is based on a predictive likelihood and this attempts to optimize performance measures such as predictive mean squared error. Thus, AICc attempts to deal with out-of-sample prediction by its very derivation. Even model {34 3*4} had an adjusted $R^2 = 0.921$. $R^2$ is a descriptive statistic and should not be used for formal model selection (it is very poor in this regard). A likelihood version of "$R^2$" is given in Appendix A and is useful when the analysis has been done in a likelihood framework.

The danger of having all the models in the set be useless arises most often in exploratory work where little thought went into hypothesizing science relationships or data collection protocols. I have seen a number of habitat–animal association models where the best model in the set had an $R^2$ value around 0.06, certainly indicating that more work is needed. In such cases, the rankings of the models carry little meaning.

Generally some assessment of the worth of the global model is suggested. This assessment might be a goodness-of-fit test, residual analysis, adjusted $R^2$, or other similar approach. If a global model fits, AICc will not select a more parsimonious model that fails to fit. Thus, it is sufficient to check the worth and fit of the global model. Often it is appropriate to provide an $R^2$ value for the best model in reports or publications.

Another approach relies on including a "null" model in the set to evaluate the worth of particular hypotheses or assumptions. Consider a study of growth in tadpoles where density is a hypothesized covariate. One could evaluate a model where growth depends on density and another model where growth is independent of density. This procedure, used carefully, might allow insights as to the worth of models in the set. Details for such evaluations are given in Chap. 4.

### 3.7.3   Prediction from the Best Model

One goal of selecting the best model is to use it in making inferences from the sample data to the population. This is model based inductive inference and prediction is one objective. In this case, prediction would come from the model structure:

$$E(Y) = \beta_0 + \beta_1(x_1) + \beta_2(x_2),$$

where $x_1$ = calcium aluminate and $x_2$ = tricalcium silicate. The least squares estimates of $\beta_1$ and $\beta_2$ allow predictions to be made from the fitted model:

$$E(\hat{Y}) = 52.6 + 1.468(x_1) + 0.662(x_2)$$

The adjusted $R^2$ for this model is 0.974 suggesting that prediction is expected to be quite good until one realizes that the out-of-sample prediction performance might be poor with a sample size of 13 and the fitting of four parameters. This issue will be further addressed in Chap. 5.

## 3.8 Ranking the Models of Bovine Tuberculosis in Ferrets

The computation of AICc from the tuberculosis data allows a ranking of the five science hypotheses and these are shown in the table below:

| Hypotheses | $K$ | $\log(\mathcal{L})$ | AICc | Rank |
|---|---|---|---|---|
| $H_1$ | 6 | −70.44 | 154.4 | 4 |
| $H_2$ | 6 | −986.86 | 1,987.2 | 5 |
| $H_3$ | 6 | −64.27 | 142.1 | 3 |
| $H_4$ | 6 | −45.02 | 103.6 | 1 |
| $H_5$ | 6 | −46.20 | 105.9 | 2 |

AICc for the model corresponding to $H_5$ is computed as

$$\text{AICc} = -2\log(\mathcal{L}(\hat{\theta})) + 2K + \frac{2K(K+1)}{n-K-1} = -2(-46.20) + 2 \cdot 6$$
$$+ (2 \cdot 6 \cdot 7)/(62-5) = 92.4 + 12 + 1.474 = 105.9.$$

Because $K = 6$ for all five models in this example, AIC and AICc would select the same model.

Parameter estimates for these models were MLEs and there were no estimates of residual variance $\hat{\sigma}^2$; instead, the maximized value of the log-likelihood was available directly. Here it is easy to compute AICc, given the number of estimable parameters (six for each model here), the sample size ($n = 62$), and the value of the maximized log-likelihood function (tabled above) for each of the five models. The reader is asked to verify the computation for a few entries in the table to be sure they understand the issues. Note, too, because a likelihood approach was used here, there is no statistics strictly analogous to an $R^2$ value in the usual (i.e., least squares) sense (but see Nagelkerke (1991) for a useful analog. A final technical note is that there is often no unique measure of "sample size" for binomial outcomes such as these; Caley and Hone were conservative in using $n = 62$ in this case.

Empirical support favors $H_4$, the dietary-related hypothesis as the best of the five hypotheses. Ranking hypotheses from best to worst was $H_4$, $H_5$, $H_3$, $H_1$, and $H_2$. Clearly, $H_2$ (transmission during mating and fighting from the age of 10 months when the breeding season starts) seems very poor relative to the other hypotheses. Such observations and interpretations will be made more rigorous in Chap. 4. At this time it might be reasonable to begin to wonder about the ranking of hypotheses if a different data set of the same size was available for analysis; would the rankings be the same? This issue is termed "model selection uncertainty" and will turn out to be very important.

## 3.9    Other Important Issues

### 3.9.1    Takeuchi's Information Criterion

I mention Takeuchi's information criterion (TIC) but it is rarely used in practice. However, it is important in the general understanding of information criteria. Akaike's derivation assumed, at one step, an expectation over the model (not full reality). This has lead to the assumption that AIC was based on a true model being in the set; although Akaike clearly stated otherwise in his papers.

Takeuchi (1976), in a little known paper in Japanese, made the first published derivation clearly taking all expectations with respect to full reality. Takeuchi's TIC is an asymptotically unbiased estimate of expected K–L and does not rest in any way on the assumption that a "true model" is in the set. TIC is much more complicated to compute than AICc because its bias adjustment term involves the estimation of the elements of two $K \times K$ matrices of first and second partial derivatives, $J(\theta)$ and $I(\theta)$, the inversion of the matrix $I(\theta)$, and then the matrix product. TIC is defined as

$$\text{TIC} = -2\log(\mathcal{L}(\hat{\theta})|\text{data}) + 2\text{tr}(J(\theta)I(\theta)^{-1}),$$

where "tr" is the matrix trace operator. Unless sample size is *very* large, the estimate of $\text{tr}(J(\theta)I(\theta)^{-1})$ is often numerically unstable; thus, its practical application is nil (I have never seen TIC used in application). However, it turns out that a very good estimate of this messy term is merely $K$ or $K + K(K + 1)/(n - K - 1)$, corresponding to AIC and AICc. Thus, it can be seen that AIC and AICc represent a *parsimonious* approach to bias correction! That is, rather than trying to compute estimates of all the elements in two $K$ by $K$ matrices, inverting one, multiplying the two, and computing the matrix trace, just use $K$ or $K + K(K + 1)/(n - K - 1)$, as these are far more stable and easy to use. [In fact, if $f$ was assumed to be in the set of candidate models, then for that model $tr(J(\theta)I(\theta)^{-1}) \equiv K$. If the set of candidate models includes any decent models, then $\text{tr}(J(\theta)I(\theta)^{-1})$ is approximately $K$ for those models.]

It is important to realize that the deviance term nearly always dwarfs the "penalty" term in AICc or TIC. Thus, poor fitting models have a relatively large deviance and, thus, the exact value of the penalty term is not critical in many cases.

### 3.9.2   Problems When Evaluating Too Many Candidate Models

A common mistake is to focus on models without full consideration of the all important science hypotheses. Armed with too little science thinking and computer software that allows "all possible" models to be fit to a hapless data set, one is ready to find a wide variety of effects that are spurious. This is a subtle but important point and there is a large statistical literature on this matter. The entire fabric of the investigation breaks down in many exploratory studies where sample size might be only 35–80 and there are 15–20 explanatory variables, leading to about 33,000 or 1,050,000 models, respectively. In these cases, one may expect substantial overfitting and the finding of many effects that are actually spurious (Freedman 1983; Flack and Chang 1987; Anderson 2001). One useful rule of thumb is when the sample size is smaller than the number of models (i.e., $n < R$), then the analysis must be viewed as only exploratory (see Burnham and Anderson 2002:267–284). If one thinks as Chamberlin suggested, the focus will be on the science issues and multiple working hypotheses. Then develop models to represent these hypotheses, keeping an eye on the science and less so on countless models that can be run easily by sophisticated (but unthinking!) software. Good application can expect $n \gg R$.

Hoeting et al. (2006) provide an example of geostatistical modeling of whiptail lizards in southern California. There were 37 predictor variables available, leading to $1.4 \times 10^{11}$ possible models. They were able to reduce the number of variables to six which resulted in a tractable 160 models. Three of these were judged to be good models and involved similar variables.

### 3.9.3   The Parameter Count K and Parameters that Cannot be Uniquely Estimated

Often there are some parameters in a model that are not uniquely estimable from the data and these should not both be counted in $K$. Such "nonidentifiability" can arise due to inherent confounding (e.g., the estimators of survival and sampling probabilities, $S_{t-1}$ and $f_t$, respectively, in certain band recovery models of Brownie et al. 1985). In such cases, the correct value of $K$ counts the product $S_{t-1} f_t$ as a single parameter (not two parameters). Here, it is the *estimators* $\hat{S}_{t-1}$ and $\hat{f}_t$ that are confounded, not the parameters themselves.

Smith et al. (2005) provide another example of nonidentifiablity in their study of entomological inoculation rates and *Plasmodium falcipraum* infection in children in Africa. Their best model was

$$PR = 1 - \left(1 + \frac{b\varepsilon}{rk}\right)^{-k},$$

where PR = parasite ratio, $b$ = transmission efficiency, $\varepsilon$ = annual entomological inoculation rate, $r$ = inverse of the expected time to clear an infection and $1/k$ = the coefficient of variation of the population infection rate. They found that $b$ and $r$ were exactly collinear, only the ratio $b/k$ was relevant or identifiable. Thus, $K$ would be 3 in this case: $(b/r)$, $\varepsilon$, and $1/k$. If an error distribution was included, then $K$ would be increased by 1. Nonestimability and nonidentifiability are common issues in some life science problems.

Sometimes a parameter is estimated on a boundary and this can be confusing. If the parameter being estimated is a probability (e.g., a transition probability of moving from state $i$ to state $j$, say $\psi_{ij}$), then it may be that the MLE $\hat{\psi}$ is either 0 or 1 (i.e., on a boundary). Often the estimated standard error is 0, with a confidence interval of 0 width. In such cases, this parameter estimate must enter the count for $K$, even though some software may indicate such a parameter was "not estimated." Here, the parameter was estimated, it just happened that the most likely estimate was 0 or 1 (a boundary) and it should be counted in $K$.

Another technical point is the case where the iterative numerical procedure fails to "converge" in likelihood-based estimation. This condition is important and is nearly always noted on the output by the software. Until convergence is obtained or the specific situation understood, the analysis for that model should not go forward (i.e., the maximum of the log-likelihood function has not been found). Often, the failure to converge is due to the log-likelihood surface (see Appendix A) being nearly perfectly flat over some region in the parameter space. Thus, repeated tries to find the exact maximum point can fail. Alternatively, the log-likelihood surface might have more than a single mode, making valid inference more difficult (but there are many ways to address this problem).

### 3.9.4    Cross Validation and AICc

Basing AICc on the expectation (over $\hat{\theta}$) of $E_x[\log(g(x|\hat{\theta}(y)))]$ provides the criterion with a cross validation property for independent and identically distributed samples (Stone 1974, 1977). Golub et al. (1979) show that AIC asymptotically coincides with generalized cross validation in subset regression (also see review by Atilgan 1996). These are important results for application and are another by-product of Akaike's predictive likelihood. The practical utility of these findings suggest that computer-intensive cross validation results will average about the same result as just using AICc.

### 3.9.5    Science Advances as the Hypothesis Set Evolves

Evolution importantly involves time and information. Consider an investigator with $R = 5$ good, plausible science hypotheses, a mathematical model representing each of the five, and a set of relevant data from a proper collection scheme. Upon completion of the analysis using an information-theoretic approach, it may become clear that two of the hypotheses have virtually no empirical support; their *likelihoods* (Chap. 4) are perhaps 3,000 or 6,600 to one of having utility.

At this point, one wants the hypothesis set to "evolve" allowing rapid progress in learning and understanding the system under study. First, the set is now reduced to three plausible alternatives (i.e., the two hypotheses lacking empirical support can be dropped from further consideration). Second, perhaps the three remaining hypotheses can be refined or their models can be made a better reflection of the intended hypothesis. Third, more hard thinking and consideration might lead to the introduction of one or more new hypotheses into the set. At this point, new data are collected and the process is repeated.

There is some art involved in this evolution. For example, if a large amount of new data can be anticipated, one must be careful and not discard some intricate hypotheses with high dimensioned models because such models might find support with a much larger data set. Often a scientist might prefer a more simple model if it predicts well, has parameters that are directly related to the system, and captures the main effects. Thus, there is some flexibility to use a model other than that estimated to be best for some inferences. An important aspect of science is that it never stops; each step (the set continually evolves) tends to lead to new and better understanding. Some steps might go "backward" for awhile, but science has a way of correcting missteps.

## 3.10   Summary

The crucial, initial starting point for advancement in the life sciences is a set of "multiple working hypotheses" defined prior to data analysis. These are the result of a determination to address the background science of the issue at hand. Following this important step, the science of the matter, experience, and expertise are used to define an *a priori* set of candidate models, representing the hypotheses. **These are important philosophical issues that must receive increased attention.** The research problem should be carefully stated, followed by careful planning concerning the sampling or experimental design. Sample size and other planning issues should be considered fully before the data gathering program begins. Information-theoretic procedures are not for rectifying poor science questions or resurrecting bad data.

Of course, hypotheses and models not in the set remain out of consideration. AICc can be useful in selecting the best model in the set; however, if all the models are very poor, AICc will still select the one estimated to be best and rank the rest. However, even that relatively best model will be poor in an absolute sense. Thus, every effort must be made to assure that the set of hypotheses and models is well founded.

A good model separates "information" from noise or noninformation. We are not trying to model the data; instead we are trying to model the information in the data. We are trying to use the data at hand to make inferences about the process that generated the data and to make good out-of-sample predictions.

The underlying basis of AIC is (heuristically) a model that minimizes

$$E_{\hat{\theta}}(I(f, g(\cdot \mid \hat{\theta}))).$$

This is the K–L distance or information loss, given the model is fit to the data (in the sense that parameters are estimated from the data). When faced with data and unknown model parameters, the target changes to *expected* K–L information and is based on the fitted model.

The Principle of Parsimony provides a conceptual guide to model selection, while expected K–L information provides an objective criterion, based on a deep theoretical justification. AICc provides a practical method for model selection and associated data analysis and are estimates of expected K–L information. AIC, AICc, and TIC represent extensions of classical likelihood theory, are applicable across a very wide range of scientific questions, and AICc is quite simple to use in practice.

I advise that the theories underlying the information theoretic approaches and hypothesis testing are fundamentally quite different. AICc is not a "test" in any sense and there are no associated concepts such as test power or $\alpha$-levels; statistical hypothesis testing represents a very different paradigm. The results of model selection under the two approaches might happen to be similar with simple problems and a large amount of data; however, in more complex situations, with many candidate models and less data, the results can be quite different.

## 3.11   Remarks

Guiasu (1977) and Cover and Thomas (1991) provide an overview of the broad field of information theory for those wanting to read more. Akaike's main results on this issue appeared in 1973, 1974, and 1977, but these are for the statistically and mathematically gifted. His broader and more contextual papers appeared in 1981a and b, 1985, 1992, and 1994 and these are more readable by mortals. Many of Akaike's collected works were published by Parzen et al. (1998) and insights into Akaike's career are found in Findley and Parzen (1995).

Cohen and Thirring (1973) and Broda (1983) give a full account of Boltzmann's life and science contributions. It is said that Boltzmann was the nineteenth century's greatest scientist. Gallager (2001) and Golomb et al. (2002) provide information on Claude Shannon's life and contributions to information theory. It is said that Shannon's Master of Science thesis is the most famous or well-known thesis ever written. Claude Shannon wanted to go into genetics and his Ph.D. dissertation (never published) was on genetics. Like Boltzmann, Shannon was working far beyond existing science frontiers of the time.

Pronunciation is important; Akaike is pronounced with an accent on the "ka" and the "i" is pronounced like an "e" – AKAeke. Leibler is pronounced with the accent on the "i" while the "e" is silent – LIbler.

Akaike (1973) considered his information criterion to be a natural extension of R. A. Fisher's likelihood theory. It is of historic interest that Fisher (1936) anticipated such an advance when he wrote,

*"an even wider type of inductive argument may some day be developed, which shall discuss methods of assigning from the data the functional form of the population."*

Zellner's book (Zellner et al. 2001) and particularly Forster's chapter make for interesting reading about modeling and inference (also see Jessop 1995 and Wallace 2004). Some authors view $K$, the asymptotic bias correction term in AIC, as a measure of "complexity." Perhaps no harm is done in viewing it this way; however, it does not need to be so defined. I doubt if our word "complexity" can be quantified in a satisfactory way as a single number or quantity. I view $K$ as merely an asymptotic bias correction term.

A technical point: Given a parametric structural model, there is a unique value of $\theta$ that, in fact, minimizes K–L information $I(f, g)$. This (unknown) minimizing value of the parameter depends on truth $f$, the model $g$ through its structure, the parameter space, and the sample space (i.e., the structure and nature of the data that can be collected). In this-sense there is a "true" value of $\theta$ underling ML estimation (let this value be $\theta_o$). Then $\theta_o$ is the absolute best value of $\theta$ for model $g$; actual K–L information loss is minimized at $\theta_o$. If one somehow knew that model $g$ was, in fact, the K–L best model, then the MLE $\hat{\theta}$ would estimate $\theta_o$. This property of the model $g(x|\theta_o)$ as the minimizer of K–L, over all possible $\theta$, is an important feature involved in the derivation of AIC or AICc (Burnham and Anderson 2002:Chap. 7).

Another technical point concerns $f$ the conceptual full reality. At a high level of abstraction we consider entities such as random variables and probability distributions. These are intellectual ways of thinking and understanding. Such abstraction carries over the notion of full reality which I denote as $f$. This symbol relates to the *concept* of the best "model" of full reality. There are no unknown parameters; reality may not even be parameterized. We parameterize models in an effort to understand full reality, $f$.

Some computer software use the expression $2\log(\mathcal{L}) - 2K$ as "AIC" and then the objective is to maximize this across models. While this is not incorrect, it is certainly confusing and thus statements such as "bigger is better" must be displayed to help the user from getting the worst model and thinking it is the best model! I recommend against this practice; AIC has a clear definition and I think it is best to use it.

A colleague wrote his explanation for the "pretending variable" issue. Consider two models, (1) $E(Y) = \beta_o + \beta_1(X_1)$ and (2) $E(Y) = \beta_o + \beta(X_1) + \beta_2$ (independent random variable). Both models will have essentially the same deviance because of the addition of only a "noise" variable. The models will differ by one parameter; the first model has $K = 3$, whereas the second model has $K = 4$, hence, $\Delta$ for the second model will be bigger that the first model by two. The clue here is that the deviance did not change with the addition of another variable and its parameter.

The nonparametric bootstrap can be used in model selection; this was investigated by Burnham and Anderson (2002) and in general we found the performance of the analytic approach to be as good as if not better. Given the computer-intensive nature of the bootstrap, we have not given this approach much more attention or interest. Still, it is a general approach and might find use in some cases.

One can find in the published literature that AIC is only for nested models; this statement is incorrect. Likewise, other literature states that AIC is only for nonnested models. This, too, is incorrect. The general derivation (e.g., Takeuchi 1976 or Burnham and Anderson 2002:Chap. 7) makes no restriction concerning nestedness.

The methods outlined in this book apply to virtually all problems where a likelihood exists (Lahiri 2001). In addition, there are general information-theoretic approaches for models well outside the likelihood framework (Qin and Lawless 1994; Ishiguro et al. 1997; Hurvich et al. 1998; Pan 2001a,b). There are now model selection methods for generalized estimation equations, kernel methods, martingales, nonparametric regression, and splines. Thus, methods exist for nearly all classes of models we might expect to see in the theoretical or applied life sciences.

Richard Leibler explained to me (about 1997) that many people thought their 1951 paper was a direct result from the war effort. Instead, the motivation for that (now) famous paper was to provide a rigorous definition of what Fisher meant by the word "information" in relation to his "sufficient statistics." Indeed, they showed that all the "information" in the data was contained in sufficient statistics, given the model; just as Fisher had alleged. Few people realized the importance of the 1951 paper; they got no reprint requests for their paper for many years! Also interesting, Leibler had never realized that K–L information was the negative of Boltzmann's entropy.

The K–L information or distance has also been called the K–L discrepancy, divergence, and number – I will treat these terms as synonyms, but tend to use *information* or *distance* in the material here (see Ullah 1996 for applications). Later, Kullback (1987) preferred the term *discrimination information*. Kullback served as head of the Statistics Department at George Washington University from 1964–1972 where he had a profound impact. He believed that information theory provides a unification of known results, leads to generalizations and the derivation of new results, and offers a unifying principle in statistics.

The second-order bias correction (leading to AICc) stems from Suguira's (1978) work and several follow-up papers by Hurvich et al. While these papers are not theoretical contributions on the same scale as Akaike's papers, they are very important in application. One should not use AIC in standard application; people should be using AICc, the second-order version of AIC (or derive new results if a specific distribution is required, see Burnham and Anderson 2002:Sect. 7.4.2).

It must be noted that Rissanen (1989, 1996) has derived a sophisticated model selection theory based on information and coding theory. His approach is very different, both conceptually and mathematically, than that presented in this book. His initial contribution was MDL for minimum description length and he has extended this in later publications and books (Rissanen 2007). The MDL approach does not require prior distributions on parameters or models and many people would see this as an advantage in science issues. The MDL result was the same form as BIC (Appendix D), but later theory expands on this result. I will not go further into Rissanen's work as it is quite technical unless one has the required background in coding theory. I note only that this interesting class of "information-theoretic" alternatives exists.

Akaike (1973, 1974) used what he called a predictive log-likelihood in deriving his information criterion; this has advantages and properties that are still not well recognized in the literature. Full discussion of his approach is technical and I will not provide more than a few insights here (see Akaike 1973, 1987:319, 1992; Bozdogan 1987; Sakamoto 1991; deLeeuw 1992; and Burnham and Anderson 2002:Chap. 7). His approach involves a statistical expectation based on a different, independent sample. It is this second expectation over a conceptually independent "data set" that provides AIC with a cross validation property (see Tong 1994; Stone 1977). Akaike's predictive log-likelihood is

$$E_p[\log(\mathcal{L}(\hat{\theta}))] = E_f E_f[\log(\mathcal{L}(\hat{\theta}_y)|x)].$$

Thus, $E_f E_f[\log(\mathcal{L}(\hat{\theta}_y)|x)]$ is the "target" of estimation; under certain conditions, $\log(\mathcal{L}(\hat{\theta})) - K$ is an estimator of this target when sample size is large (asymptotically). The expectation over both the data $x$ and the estimated parameters $\hat{\theta}$ are taken with respect to the true $f(x)$. This expectation addresses the technical issue of parameter uncertainty. Zucchini (2000) provides a nice introduction to model selection using a well chosen example that helps understanding. Konishi and Kitagawa (2007) provide a technical review of these issues and introduce another extension.

It is not easy to see why including a great many models in the candidate set is poor practice. One sidesteps this issue if they concentrate on science hypotheses first, and then think hard about a good model to represent each hypothesis. It is the availability of software to "run practically everything in sight" that leads to this confusing issue. Zucchini (2000) provides several figures to illustrate the dangers of evaluating an excessive number of models.

Some software packages offer a "stepwise AIC" as an option for model selection (often termed variable selection in regression analysis). This is hardly in the spirit of the information-theoretic approach and I strongly recommend against it. Such *ad hoc* procedures strongly encourage an "all possible models" approach that seems counter to good science. Good science has to be more about hard thinking and developing what seem to be plausible hypotheses; then proceed to build models as a way to evaluate the strength of evidence for these *a priori* hypotheses.

## Shannon Entropy

Claude Shannon, working during the 1940s, is often regarded as the father of information theory. Shannon (1948) justified entropy for discrete variables with discrete and finitely many outcomes as

$$H = -\sum_i P_i \log P_i,$$

where $P_i$ is the probability of outcome $i$. He approached this by positing three conditions that information (in the context of probability) should satisfy. He then proved that $H$ was the unique solution that satisfied the conditions. The entropy of a probability distribution is

$$H = -\int p(x) \log p(x) dx,$$

where $p(x)$ denotes the probability density with respect to the measure $dx$. Ecologists toyed with computing entropies in the early 1970s (an endeavor that Shannon termed the "bandwagon" in an editorial in the *Transactions of Information Theory*). While hundreds of papers presented entropies in leading ecological journals, most people now believe that this avenue produced little of value. In actuality, the definition of information was designed to help communication engineers send messages, rather than to help people understand the *meaning* of messages.

Goldman (1953) considers information to be the *difference* between our uncertainty before and after receiving a message. In this thinking, information is not an absolute quantity as implied from $H$, but is seen as a *change* in uncertainty. Let $q_i$ be the probability of the $i$th event before receiving the message and $p_i$ be the revised probability after receipt of the message. The change in the uncertainty is

$$[-\log(q_i)] - [-\log(p_i)] = \log(p_i) - \log(q_i) = \log(p_i/q_i).$$

If the message received indicates that the $i$th event is certain, then $p_i = 1$ and $\log(p_i) = 0$, resulting in a change in information of $-\log(q_i)$. Jessop (1995) terms this "surprisal." Taking the expectation

$$E[\log(p_i/q_i)] = \sum_i p_i \log(p_i/q_i)$$

and is the discrete version of K–L information! Kullback–Leibler information
is an extension of Shannon's contribution and is sometimes called a "relative
entropy" (Hobson and Cheng (1973)). The K–L information between models
(probability distributions) is a *fundamental quantity* in science and information theory and is the logical basis for model selection.

## Boltzmann's Entropy

Ludwig Boltzmann, working in the late 1800s, originally defined entropy in
thermodynamics, demonstrated the second law of thermodynamics (e.g., there
could not be a perpetual motion machine), and proved the irreversibility of
entropy. Entropy is "disorder," max entropy is maximum disorder or minimum
information. While the theory of entropy is a large subject by itself, readers
here can think of entropy as nearly synonymous with uncertainty.

Conceptually, Boltzmann's entropy is $-\log(f(x)/g(x))$ and taking its expectation one gets

$$E_f\left(-\log\left(\frac{f(x)}{g(x)}\right)\right) = \int f(x)\log\left(\frac{f(x)}{g(x)}\right)dx,$$

which is K–L information (see Good 1979). It is fascinating that Kullback–Leibler information is equal to the negative of Ludwig Boltzmann's
entropy. Thus, minimizing the K–L information or distance is equivalent
to maximizing the entropy; hence the name *maximum entropy principle*
(Jaynes 1957).

Maximizing entropy is subject to a constraint – the model of the information
in the data. A good model contains the information in the data, leaving only
"noise." It is the noise (entropy or uncertainty) that is maximized under the
concept of the *Entropy Maximization Principle*. Minimizing K–L information then results in an approximating model that loses a minimum amount of
information in the data. Entropy maximization results in a model that maximizes the uncertainty, leaving only information (the model) "maximally"
justified by the data. The concepts are equivalent, but minimizing K–L distance (or information loss) certainly seems the more direct approach. In
summary,

$$- \text{entropy} = K - L \text{ information}$$

and K–L information is often referred to as negative entropy or negentropy.

Boltzmann's discoveries concerning entropy are seen as the zenith of nineteenth century science. Of course, K–L information was derived along very
different lines than entropy; the mutual convergence is striking and suggests
something very fundamental. K–L information is *averaged* entropy, hence

the expectation with respect to $f$. Then, $-E(\text{entropy}) = \text{K--L information}$. Boltzmann derived the fundamental theorem that,

**entropy is proportional to log(probability).**

Entropy, information, and probability are thus linked, allowing probabilities to be multiplicative while information and entropies are additive.

## 3.12   Exercises

1. Cox (2006:117) states, "The relevance of automatic model selection depends strongly on the objectives of the analysis, for example as to whether it is for explanation or for empirical prediction." By "automatic model selection" I think he means criteria such as AICc, BIC, TIC, etc. Can examples be found where an investigator might need to use, say, AICc for prediction, but another criteria (or an entirely different approach?) for explanation (given the data are fixed)? What theory might bear on his statement? What practical advice might be given as to how to approach model selection when the main objectives of the analysis might vary? Discuss this with colleagues and see if the premise has merit.

2. In ecology increased diversity is often associated with ecotones. In a sense, Akaike was at a science ecotone when he saw a way to relate information theory and statistical theory in his AIC. Can you think of other parallels of this nature? What might this say about coursework to be taken by an exceptional Ph.D. student?

3. Akaike found an analytic expression for the asymptotic bias when the maximized $\log(\mathcal{L})$ was used as an estimator of expected K--L information; this bias correction was simply $K$, the number of estimated parameters in the model. Give other examples of estimators in your field where bias adjustments have been found.

4. AICc is simple to compute and understand, but it rests on very deep statistical theory. This makes it an ideal science tool. Give other examples where this is the case.

5. The data on hardening of Portland cement had four predictor variables; this leads to $2^4 - 1 = 15$ models. If all 2- and 3-way interactions would have been added, how many models would there be? What is the danger here in focusing on the models during data analysis?

6. Traditional statistics provided judgments about "significance" and this is related to some predefined, but arbitrary $\alpha$-level. Such terms and dichotomies are shunned under the information-theoretic approach. Discuss and attempt to reconcile your thoughts on this matter of fixed dichotomies.

7. Examine a recent issue of a journal in your field of interest. Can you find a well written paper that carefully sets out several working hypotheses before data analysis? In some subdisciplines, such papers can be easily found. Once having found such a paper, what approach did the authors use as a measure of "strength of evidence" for and against the science hypotheses?

8. Atmar (2001) wrote a fitting obituary of Claude Shannon that makes interesting reading. He also references Dawkins (1986:111–112):

> A few years ago, if you asked almost any biologist what was special about living things as opposed to nonliving things, he would have told you about a special substance called protoplasm. Protoplasm wasn't like any other substance; it was vital, vibrant, throbbing, pulsating, "irritable" (a schoolmarmish way of saying responsive).… When I was a school boy, elderly textbook authors still wrote of protoplasm, although, by then, they really should have known better. Nowadays you never hear or see the word. It is as dead as phlogiston and the universal aether. There is nothing special about the substances from which living things are made. Living things are collections of molecules, like everything else. What is special is that these molecules are put together in much more complicated patterns than the molecules of nonliving things, and this putting together is done by following programs, sets of instructions for how to develop, which the organisms carry around inside themselves. Maybe they do vibrate and throb and pulsate with "irritability," and glow with living warmth, but these properties all emerge incidentally. What lies at the heart of every living thing is not a fire, not a warm breath, not a "spark of life." It is information, words, instructions. It you want a metaphor, don't think of fires and sparks and breath. Think, instead, of a billion discrete, digital characters carved in tablets of crystal. If you want to understand life, don't think about vibrant, throbbing gels and oozes, think about information technology.

This thinking is certainly exciting – evolution and life are about information! Think hard about this and discuss it with colleagues and instructors. Is evolution so much about information? Where might these concepts lead us in the life sciences?

9. Assume you have some data on a well-defined science issue and the models for the four hypotheses are complimentary log–log models for a binary response variable. You have $n = 19$ and the global model has $K = 6$ parameters and AIC has been used as the first step in providing measures of strength of evidence for the four hypotheses. What is the issue that might be of concern here? Why?

10. Your new student questions the concern about models with "too many" parameters that must be estimated from the data. You speak of overfitting but he insists that biology is complex and some simple models are not "realistic." Prepare a clear response to help him understand this issue.

11. Recompute the information at the beginning of Sect. 3.7 using AIC. Provide your interpretation of any differences you encounter. What is the "moral" of this example?

12. You have just been hired by a government laboratory that has access to a very large amount of data from a Superfund site in Georgia. The questions were well formed, data collection was quite sophisticated, and sample sizes were very large by any usual standard. You are to work in a team situation and the team members have been educated and experienced in a variety of relevant disciplines. Some members of the team want to do an analysis using AIC, while others have heard about TIC and they favor this approach. They look to you for advice and council. What do you tell them? Why?

13. The bovine tuberculosis study by Caley and Hone (Sect. 3.8) is interesting in many ways. For example, they collected data by gender (also across five sites) and gender was a variable in all their models. A reviewer with expertise in *mustelids* claims that gender is unimportant in disease transmission and should not have been in the models (for parsimony reasons, if no other). Using AICc, how could you determine if the deletion of gender was better than models including gender? Be specific but concise.

# 4

# Quantifying the Evidence About Science Hypotheses

*Richard Arthur Leibler* (1914–2003) was born in Chicago, Illinois on March 18, 1914. He received a Bachelors and Masters degree in mathematics from Northwestern University and a Ph.D. in mathematics at the University of Illinois (1939). After serving in the Navy during the war, he was a member of the Institute for Advanced Study at Princeton and a member of the von Neumann Computer Project 1946–1948. From 1948–1980 he worked for the National Security Agency (1948–1958 and 1977–1980) and the Communications Research Division of the Institute for Defense Analysis (1958–1977). He then was the president of Data Handling Inc., a consulting firm for the Intelligence Community. He received many awards, including the Exceptional Civilian Service Award.

The ability to simply rank science hypotheses and their models is a major advance over what can be done using null hypothesis tests. However, much more can be done, all under the framework of "strength of evidence," for hypotheses in the *a priori* candidate set. Such evidence is exactly what Platt (1964) wanted in his well-known paper on *strong inference*. I begin by describing four new evidential quantities.

## 4.1    $\Delta_i$ Values and Ranking

In Chap. 2 it became clear that AIC values were relative rather than absolute
for three reasons: (1) sample size impacted the size of AIC values, (2) there
was the unknown constant $E_f[\log(f(x))]$ in the derivation of AIC from K–L
information, and (3) some terms in the model set that are constant across models are often omitted. Simple differencing renders these issues moot. These
differences, denoted as $\Delta_i$, are standardized by the AICc value for the best
model (the minimum AICc value). In fact, such differencing defines the best
model as always having $\Delta_{best} \equiv 0$.

---

### AICc Differences are Fundamental Units

Formally, the differences, $\Delta_i$, are defined as

$$\Delta_i = AICc_i - AICc_{min}.$$

These values are estimates of the expected K–L information (or distance)
between the best (selected) model and the $i$th model. These differences
apply when using AICc, QAICc (Sect. 6.2), or TIC, are on the scale of
information, and are additive.

---

At this point we have science hypotheses and their associated models on a standard measurement scale. Although a scale of "information" might seem odd at first, it
is little different than working with meters and kilometers or feet and miles.

Kullback–Leibler information is the distance from each of the models to
full reality, whereas the $\Delta_i$ values relate to the distance between each of the
models to the best one (Fig. 4.1). Everything is scaled to the best model where
$\Delta_{best} \equiv 0$. This is convenient and is like so many other things in our experience. For example, in horse racing everything is scaled to the winner (quickest
horse). The absolute time of the winning horse is unimportant because track
conditions change from race to race and year to year. So, we speak of the
winning horse and the second horse being two lengths behind, etc. Putting
various science hypotheses in terms of the best one (i.e., the one with the most
empirical support) is expected and should not be mistaken as arbitrary.

In a similar way, we measure the height of mountains and cities as the distance above sea level. Sea level has been convenient as a basis in scaling heights,
much like $\Delta_{best}$ is convenient in scaling information and assessing the distance to
other hypotheses and their models. Of course, such scaling to $\Delta_i$ values does not
change the ranks based on AICc. It does make the examination of ranks visu-
ally easy as one merely looks for the model with $\Delta = 0$ and realizes that this is
the model estimated to be the best (closest to truth). We must bear in mind that
these are estimates and if we had a replicate data set of the same size and from
the same process, a different hypothesis might be estimated to be the best in
that case. We will quantify this uncertainty (called *model selection uncertainty*)
using simple methods outlined in this chapter and Chap. 5.

"DISTANCES"

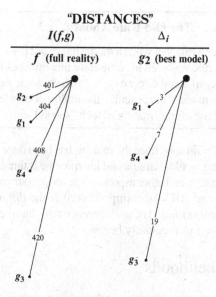

FIG. 4.1.  Kullback–Leibler distances are with respect to the conceptual full reality (*left*), while the $\Delta_i$ are distances with respect to the best model and are on the scale of information.

An interesting issue arises because the $\Delta_i$ are on an information scale (Sect. 3.3.5). Under some fairly weak assumptions it turns out that only science hypotheses and their models having $\Delta_i$ values in the range of 0 to perhaps 9–12 are plausible to most objective people (plausibility is a value judgment!). Even though AICc values for a particular problem might be in the 175,800–179,400 range, only models where $\Delta$ is within about 0 to 9 or 12, or 14 have much credibility. This will turn out to be a very useful result in practical application with real data. When I give the "window" above, I am deliberately trying to be vague about the upper bound as I do *not* want readers to consider some arbitrary cutoff (as in the $\alpha$-level in testing theory). Science is about estimation and understanding; it is not about cutoffs or dichotomies.

Still, models with $\Delta$ values close to 0 have a lot of empirical support. Models with $\Delta$ values in the rough range 4–7 have considerably less support, whereas models with $\Delta$ values in the fringes (say 9–14) have relatively little support. Others, still further away, might be dismissed by most observers as *implausible*. The rationale for these rough guidelines is provided in Sect. 4.4; in addition, Royall (1997) offers similar guidelines. If observations are not independent but are assumed to be independent then these simple guidelines cannot be expected to hold. Likewise, if there are thousands of models, these guidelines may not hold entirely (however, if there are thousands or millions of models, then the endeavor is questionable anyway). The reader should not take these guidelines as inviolate as there are situations to which they may not apply well. Approaches to allow a more careful interpretation of the evidence are covered in the following material; thus, these rough guidelines are not necessary.

---

**The Old Rule About $\Delta_i > 2$**

One occasionally sees a rule that when a model has $\Delta_i > 2$ it is a poor model, or of little value, etc. This rule was seen in some early literature but I advise against it. Models with $\Delta_i$ values of 2 or 4 or even 6–7 or so have some meaningful support and should not be dismissed. In particular, the multimodel inference framework (Chap. 5) invites the use of information in such "second string" models.

---

The analysis is hardly finished once the ranking has been done and the best model has been estimated. One would examine and interpret the estimates of model parameters, the covariance matrix, and other aspects of the model estimated to be the best.

The absolute value of AICc is unimportant, it is the *differences* that can be directly related to *information*. These differences are an important quantity in the methods introduced immediately below.

## 4.2 Model Likelihoods

Parameter estimation under a likelihood framework is based on

$$\mathcal{L}(\underline{\theta}|\underline{x},g_i),$$

meaning "the likelihood as a function of only the unknown parameters ($\theta$), *given* the data ($\underline{x}$) and the particular model ($g_i$, such as binomial or normal or log-normal)." The likelihood is a function of the unknown parameters that must be estimated from the data (Appendix A). Still, the point is, this function allows the computation of likelihoods of various (tentative) parameter values. Likelihood values are relative and allow comparison. The objective is to find the parameter value that is most *likely* (i.e., the one that maximizes the likelihood) and use it as the MLE, the asymptotically best estimate of the unknown parameter, given the data and the model. With this background as a backdrop, I can introduce the concept of the likelihood of a model, given the data.

---

**The Likelihood of Model $i$, Given the Data**

The concept of the likelihood of the parameters, given the data and the model, i.e., $\mathcal{L}(\theta \mid \underline{x}, g_i)$ can be extended to the likelihood of model $i$ given the data, hence $\mathcal{L}(g_i \mid \underline{x})$,

$$\mathcal{L}(g_i \mid \underline{x}) \propto \exp\left(-\tfrac{1}{2}\Delta_i\right).$$

Akaike suggested this simple transformation in the late 1970s. The likelihood of a model, given the data, offers the analyst a powerful metric in assessing the strength of evidence between *any* two competing hypotheses. This likelihood is very different from the usual one used in parameter estimation (obtaining MLEs). Both likelihoods are relative and useful in comparisons; they are not probabilities in any sense.

---

This simple expression allows one to compute the discrete likelihood of model $i$ and compare that with the likelihood of other hypotheses and their models. Such quantitative evidence is central to empirical science. Chamberlin and Platt would greatly appreciate having the (relative) likelihood, based on the data, of each of his multiple working hypotheses. Notice that the $-\frac{1}{2}$ in the simple equation above merely removes the $-2$ that Akaike introduced in defining his AIC. Had he not used this multiplier, the likelihood of model $i$ would have been just $\exp(\Delta_i)$ or $e^{\Delta i}$.

Likelihoods are relative and have a simple raffle ticket interpretation. One must think of likelihoods in a stochastic sense, such as the chance of winning a raffle based on the number of tickets the opponent has. Likelihoods are not like lifting weights, where the results are largely deterministic; if someone can lift considerably more than his/her opponent, the chances are good that he/she can do this again and again. There is little or no variation or "chance" in such activities. It is useful in evaluating science hypotheses to think in terms of the number of tickets each of the $R$ hypotheses might have. If $H_3$ has a likelihood of 3 and $H_5$ has a likelihood of 300, then it is clear that evidence points fairly strongly toward support of hypothesis $H_5$ as it has 100 times the empirical support of hypothesis $H_3$ (we can say formally that it is 100 times more *likely*). Models having likelihoods of 3 and 300 are similar in principle to two people, one having three raffle tickets and the other having 300 tickets. Likelihoods are another way to quantify the strength of evidence between any model $i$ and any other model $j$; there is no analogy with the "multiple testing problem" that arises awkwardly in traditional hypothesis testing. Computation of the model likelihoods is trivial once one has the $\Delta_i$ values.

## 4.3   Model Probabilities

Before proceeding to define model probabilities, we must define a relevant target value of such probabilities. Given a set of $R$ models, representing $R$ science hypotheses, one of these models is, in fact, the best model in the K–L information or distance sense. Like a parameter, we do not know which of the models in the set is actually *the* K–L best model for the particular sample size. Given the data, the parameters, the model set, and the sample size, one such model *is* the K–L best; we do not know which model is the best but we can estimate it using AICc. Moreover, we can estimate the uncertainty about our selection (our estimate of the model that is the best). This is crucial; we need a measure of the "model selection uncertainty." The target is not any notion of a "true model," rather the target is the actual best-fitted model in an expected K–L information sense. This concept must include the uncertainty in estimating the model parameters.

It is important not to confuse the "K–L best" model with a "good model." If all the models in the set are poor, these methods attempt to identify the best of these but in the end they all remain poor. Thus, one should examine such things as adj $R^2$, residual plots, and goodness-of-fit tests to be sure some of the models in the set are worthwhile.

It must be noted that the best model in a K–L sense depends on sample size. If $n$ is small then the K–L best model will be of relatively low dimension. Conversely, if $n$ is large, the K–L best model will be richer in structure and parameterization. These concepts tie back to the Principle of Parsimony (Sect. 2.3.4) and tapering effect sizes (Sect. 2.3.5). These are not easy concepts to grasp but are fundamental to model based inference. Oddly, the mathematical calculations here are trivial relative to the more difficult conceptual issues.

---

### Model Probabilities

To better interpret the relative likelihoods of models, given the data and the set of $R$ models, one can normalize these to be a set of positive "Akaike weights," $w_i$, adding to 1,

$$w_i = \frac{\exp\left(-\tfrac{1}{2}\Delta_i\right)}{\sum\limits_{r=1}^{R} \exp\left(-\tfrac{1}{2}\Delta_r\right)}.$$

These weights are also Bayesian posterior model probabilities (under the assumption of savvy model priors) and the formula applies when using AICc, QAICc (see Sect. 6.2), or TIC. A given $w_i$ is the probability that model $i$ is the expected K–L best model

$$w_i = \text{Prob}\left\{g_i \mid \text{data}\right\}.$$

These probabilities are another weight of evidence in favor of model $i$ as being the actual K–L best model in the candidate set. These $w_i$ values are most commonly called model probabilities (given the model set and the data). These can be easily computed by hand, but a simple spreadsheet might avoid errors or too much rounding off. The term "Akaike weight" was coined because of their use in multimodel inference (Chap. 5) and before it was realized that these could be derived as Bayesian posterior model probabilities. The term *model probability* will often suffice.

---

The estimated K–L best model (let this best model be indexed by b) always has $\Delta_b \equiv 0$; hence, for that model $\exp(-(\tfrac{1}{2})\Delta_b) \equiv 1$. The odds for the $i$th model actually being the K–L best model is just $\exp(-(\tfrac{1}{2})\Delta_i)$. It is often convenient to reexpress such odds as the set of model probabilities as above.

The bigger a $\Delta_i$ is, the smaller the model probability $w_i$, and the less plausible is model $i$ as being the actual K–L best model for full reality based on the sample size used. These Akaike weights or model probabilities give us a way to calibrate or interpret the $\Delta_i$ values; these weights also have other uses and interpretations (see below). While most forms of evidence are relative, model probabilities are absolute, conditional on the model set.

## 4.4   Evidence Ratios

In Sect. 4.1, some explanation was given to help interpret the $\Delta_i$ values. Evidence ratios can be used to make interpretation more rigorous and people should *not* use these as automatic cutoff values. Evidence is continuous and arbitrary cutoff points (e.g., $\Delta_i > 3$) should not be imposed or recognized.

---

**Evidence Ratios**

Evidence ratios between hypotheses $i$ and $j$ are defined as

$$E_{i,j} = \mathcal{L}(g_i \mid x) / \mathcal{L}(g_j \mid x) = w_i / w_j.$$

Evidence ratios are relative and not conditioned on other models in or out of the model set. Evidence ratios are trivial to compute; just a ratio of the model probabilities or model likelihoods.

---

Evidence ratios can be related back to the differences, $\Delta_i$. An evidence ratio of special interest is between the estimated best model (min) and some other model $i$. Then the $\Delta$ value for that $i$th model can be related to the best model as

$$E_{min,i} = w_{min} / w_i = e^{(-(1/2)\Delta_i)}.$$

Evidence ratios also have a raffle ticket interpretation in qualifying the strength of evidence. For example, assume two models, A and B with the evidence ratio = $\mathcal{L}(g_A|data)/\mathcal{L}(g_B|data) = 39$. I might judge this evidence to be at least moderate, if not strong, support of model A. This is analogous to model A having 39 tickets while model B has only a single ticket.

Some relevant values of $\Delta_i$ and the evidence ratio are given below.

---

**Interpreting AICc Differences**

The strength of evidence for the best model vs. any other model $i$ is shown as a function of $\Delta_i$:

| $\Delta_i$ | Evidence ratio |
|---|---|
| 2 | 2.7 |
| **4** | **7.4** |
| 6 | 20.1 |
| **8** | **54.6** |
| 10 | 148.4 |
| 11 | 244.7 |
| 12 | 403.4 |
| 13 | 665.1 |
| 14 | 1,096.6 |
| 15 | 1,808.0 |
| **16** | **2,981.0** |
| 18 | 8,103.1 |
| 20 | 22,026.0 |
| 50 | 72 billion |

The second row, for example, is read as "if a model $i$ has $\Delta_i = 4$, then the best model has 7.4 times the weight of evidence relative to model $i$ (i.e., the best model has 7.4 raffle tickets whereas the other model has only one ticket). Using the information in the table above it is easy to see why models with $\Delta$ somewhere in the 8–14 range would be judged by most objective people as having little plausibility. Even odds of 55 to 1, (i.e., $\Delta = 8$) might often be judged as a "long shot." Models with delta >15–20 must surely be judged to be implausible.

A 7.4 to 1 advantage is pretty good and you might bet your used bike on a game with these odds; however, you must be careful as there is still a reasonably large chance (risk) that you would lose. Thus, a model where $\Delta = 4$ should not be dismissed; it has some reasonable empirical support. In contrast, a model with $\Delta = 16$ has but one ticket whereas the best model has almost 3,000 tickets (see table above). This is a clear case where you would not want to bet on the model where $\Delta = 16$; it is probably better to dismiss the model as being implausible! More extreme is the case where a model has $\Delta = 25$; here the odds of that model being, in fact, the best K–L model are remote (about 270,000 to 1) and most reasonable people might agree that the model should be dismissed. Still, an important point is that the evidence is the numerical value of the evidence ratio; this is where the objective science stops. Value judgments may follow and help interpret and qualify the science result. This is a good place to ask if the reader of this material is motivated to buy a ticket for a state or national lottery?

The table above makes it clear that models with $\Delta$ values below about 8 or 12 are in a window of some reasonable plausibility. Surely models with $\Delta >$ say 20 can probably be dismissed (unless the data are quite dependent (Sect. 6.2) or have been substantially compromised). No automatic cutoff is appropriate here; we must "qualify our mind to comprehend the meaning of evidence" as Leopold said in 1933. The "science answer" stops at the ranks, the model likelihoods, the model probabilities, and the evidence ratios. The *interpretation* involves a value judgment and can be made by anyone, including the investigator. Burnham and Anderson (2002:320–323) provide a more complicated example of these measures of evidence involving $T_4$ cell counts in human blood.

If the sample size is small or even of moderate size, care is needed in dismissing high-dimensional models as implausible. As sample size increases, additional effects can be identified. Often when sample size is small, there is a large amount of model selection uncertainty, keeping one from rejecting models with several parameters. This is another reason to design data collection such that sample size is as large as possible to meet objectives.

As Chamberlin pointed out, everyone wants a simple answer, even in cases where the best answer is not simple. Prior training in statistics has imprinted many of us with dichotomies that are in fact artificial and arbitrary (e.g., $P < 0.05$ in null hypothesis testing). However, in everyday life, people can live comfortably without such arbitrary cutoffs and rulings of "significance."

Consider a football score, 7 to 10. Most neutral observers would conclude the game was "close," but the team with 10 points won. No one bothers to ask if the win was "statistically significant." Of course, the winning team (hardly expected to be neutral) could claim (a value judgment) that they hammered their hapless opponents. However, alumni for the losing team might claim "last minute bad luck" or "poor refereeing" toppled their otherwise superior team. Still others might look at the number of yards rushing, the number of interceptions, and other statistics, and perhaps point to further interpretations of the evidence. In the end, perhaps all we can really say is that the game was close and if the teams played again under similar conditions, we could not easily predict the winner. This is a case where value judgments might vary widely; that is, the hard evidence is a bit thin to clearly suggest (an inference) which might be the better team.

Going further, two other teams play and the score is 3 to 35. Here, one must fairly conclude that the winning team was very dominating. The quantitative evidence is more clear in this case. Again, no issue about "statistical significance" is needed; the score (evidence) is sufficient in this case. The game was a "thumping" and any neutral observer could easily judge which team was better. If the two teams were to play again under similar conditions, one would suspect the winner could be successfully predicted (an inference). In this case, value judgments would probably vary little from person to person, based on the evidence (the score).

Summarizing, in football, the evidence is the final score and in science, evidences are things like model probabilities and evidence ratios. Interpretation of this evidence involves value judgment and these might (legitimately) vary substantially or little at all. Scientists should avoid arbitrary dichotomies and "cutoffs" in interpreting the quantitative evidence.

## 4.5   Hardening of Portland Cement

Here we return to the example of the hardening of Portland cement from Sects. 2.2.1 and 3.7 to illustrate the nature of scientific evidence

| Model | $K$ | $\hat{\sigma}^2$ | $\log \mathcal{L}$ | AICc | $\Delta_i$ | $w_i$ |
|---|---|---|---|---|---|---|
| {mean} | 2 | 208.91 | −34.72 | 71.51 | 39.1 | 0.0000 |
| {12} | 4 | 4.45 | −9.704 | 32.41 | 0.0 | 0.9364 |
| {12 1*2} | 5 | 4.40 | −9.626 | 37.82 | 5.4 | 0.0629 |
| {34} | 4 | 13.53 | −16.927 | 46.85 | 14.4 | 0.0007 |
| {34 3*4} | 5 | 12.42 | −16.376 | 51.32 | 18.9 | 0.0000 |

Readers should verify the simple computations in the last two columns of the table above. For example, using $\Delta_i = \text{AICc}_i - \text{AICc}_{\min}$, $\Delta_1 = 71.51 - \mathbf{32.41} = 39.1$, $\Delta_2 = 32.41 - \mathbf{32.41} = 0$ and $\Delta_3 = 37.82 - \mathbf{32.41} = 5.4$. The computation of the model probabilities ($w_i$) is made from

$$w_i = \frac{\exp\left(-\tfrac{1}{2}\Delta_i\right)}{\sum\limits_{r=1}^{R} \exp\left(-\tfrac{1}{2}\Delta_r\right)}$$

and this is more simple than it might appear. The first step is to tabulate the values of $\exp(-(\tfrac{1}{2})\Delta_i)$ for $i = 1, 2,\ldots, 5$ as these are needed for both numerator and denominator

$$\exp\left(-\tfrac{1}{2}\Delta_1\right)=\exp\left(-\tfrac{1}{2}\cdot 39.1\right)=0.0000,$$
$$\exp\left(-\tfrac{1}{2}\Delta_2\right)=\exp\left(-\tfrac{1}{2}\cdot 0\right)=1,$$
$$\exp\left(-\tfrac{1}{2}\Delta_3\right)=\exp\left(-\tfrac{1}{2}\cdot 5.4\right)=0.0672,$$
$$\exp\left(-\tfrac{1}{2}\Delta_4\right)=\exp\left(-\tfrac{1}{2}\cdot 14.4\right)=0.0007,$$
$$\exp\left(-\tfrac{1}{2}\Delta_5\right)=\exp\left(-\tfrac{1}{2}\cdot 18.9\right)=0.0001.$$

More decimal places should often be carried; I will give the results to four places beyond the decimal (thus, 0.0000 is not meant to reflect exactly zero, it is only zero to four places). The quantity $\sum_{r=1}^{R} \exp(-(1/2)\Delta_r)$ in the denominator is merely the sum of these five numbers, **1.068**. Finally, the model probabilities are $w_1 = 0.0000/\textbf{1.068} = 0.0000$, $w_2 = 1/\textbf{1.068} = 0.936$, $w_3 = 0.0672/\textbf{1.068} = 0.063$, and so on.

Until this chapter, we could only rank the five hypotheses and their models; this, of course, allowed the estimated best model (i.e., model {12}) to be identified. Now, we have the ability to see that two of the models can be judged to be implausible: models {mean} and {34 3∗4} (even the better of these two (i.e., model {34 3∗4}) has a model probability of only 0.00008) and might be dismissed as the set evolves. Only models {12} and {12 1∗2} have noticeable empirical support whereas model {34} has very little empirical support (model probability of 0.0007). The model probabilities are exactly what Chamberlin would have wanted, but it must be remembered that they are conditional on the model set.

What is the evidence for the interaction 1∗2? An evidence ratio answers this question: $E = 0.9364/0.0629 = 14.9$ (from the table just above). This measure indicates that the support for the model without the interaction is nearly 15 times that of the model with the interaction. What would be your value judgment, based on the evidence, in this case? In other words, how would you qualify the result concerning the interaction term?

Additional evidence here is to look at the MLE for the beta parameter ($\beta_3$) for the interaction term. We find, $\hat{\beta}_3 = 0.0042$ with an approximate 95% confidence interval of (−0.020, 0.033). One must conclude that there is little support for the 1∗2 interaction term. Notice also that the deviance (deviance

= −2 × log $\mathcal{L}$) changed little as the interaction term was added: 19.408 vs. 19.252, again making support for the interaction term dubious (this is the "pretending variable" problem, Sect. 3.6.8).

Let us imagine a member of the cement hardening team had always favored model {34} and felt strongly that it was superior to the rest. What support does the member have, based on this small sample of 13 observations? The evidence ratio of the best model vs. model {34} is 1,339 to 1: not much support of model {34}. He must quickly try to argue that the data were flawed, measurements were in error, etc. No reasonable observer will overturn odds of over 1,300:1; the evidence is strongly against the member's belief and all reasonable value judgments would confirm this.

## 4.6   Bovine Tuberculosis in Ferrets

The addition of the information-theoretic differences and model probabilities allow more evidence to be examined:

| Hypotheses | $K$ | log $\mathcal{L}$ | AICc | Rank | $\Delta_i$ | $w_i$ |
|---|---|---|---|---|---|---|
| $H_1$ | 6 | −70.44 | 154.4 | 4 | 50.8 | 0.0000 |
| $H_2$ | 6 | −986.86 | 1,987.2 | 5 | 1,883.6 | 0.0000 |
| $H_3$ | 6 | −64.27 | 142.1 | 3 | 38.5 | 0.0000 |
| $H_4$ | 6 | −45.02 | 103.6 | 1 | 0.0 | 0.7595 |
| $H_5$ | 6 | −46.20 | 105.9 | 2 | 2.3 | 0.2405 |

Here, it seems clear that the evidence strongly suggests that hypotheses 1–3 are implausible; the probability that $H_3$ is, in fact, the K–L best model is less than $4 \times 10^{-8}$ and the other two models have far less probability. This finding certainly allows the set to evolve to the next level. Support of hypotheses 4 and 5 is somewhat tied, with $H_4$ having the edge by a factor of about three times (i.e., 0.7595/0.2405 ≈ 3) the support over $H_5$. One cannot rule out the support for $H_5$ based on the evidence (Fig. 4.2).

Note that these model probabilities are conditional on the set of five hypotheses and the five model probabilities sum to 1. One can compute evidence ratios among any of the five, even if one or two of the hypotheses are deleted. Often the interest is in evidence ratios with the best model vs. some other model; however, one is free to select any models $i$ and $j$ for evaluation by simple evidence ratios. Note that because all models here have the same number of estimable parameters, the penalty term can be ignored in this particular case and one can use just the deviance (−2 × log $\mathcal{L}$) as "AICc." Finally, note that if, for some reason, model $g_2$ is dropped from the set, then the other five model results must be renormalized to sum to 1 (in this particular example it would make no difference).

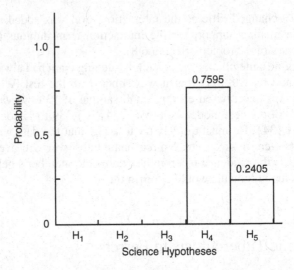

FIG. 4.2. Model probabilities for the five hypothesis concerning transmission of bovine tuberculosis in feral ferrets in New Zealand.

## 4.7   Return to Flather's Models and $R^2$

Burnham and Anderson (2002:94–96) presented a reanalysis of nine models for the species accumulation curves for data from Indiana and Ohio from Flather (1996). The science objective here was to explore the structure of the accumulation process and allow predictions concerning species accumulation and richness. Hypotheses concerning species accumulation were modeled as nonlinear regressions, and one model was found to be quite better than the other eight models. Here we will look at an interesting side issue as this gives some insights into the power of these newer methods.

The adjusted $R^2$ value in regression analysis measures the proportion of the variance in the response variable that is in common with variation in the predictor variables: it is a measure of the overall "worth" of the model, at least in terms of within-sample prediction. Adj $R^2$ for the nine models ranged from 0.624 to 0.999 (Table 4.1).

Examining Table 4.1 shows that five of the models had adj $R^2$ values above 0.980. One might infer that any of these five models must be quite good; even the worst model with $R^2 = 0.624$ might appear pretty good to many of us. Going further, two of the models (models 8 and 9) had an adj $R^2 > 0.999$; surely these models are virtually tied for the best and are both excellent models for a structural description of the accumulation process and for prediction. Although the statements above certainly seem reasonable, they are misleading. For example, the evidence ratio for the (estimated) best model (model 9) vs. the second best model (model 8) is

$$E_{min,8} = w_{min} / w_8 = w_9 / w_8 = \exp\big((\Delta_8 / 2) = \exp(163.4/2) \approx 3.0 \times 10^{35}\big)$$

TABLE 4.1. Summary of nine *a priori* models of avian species-accumulation curves from Flather (1992, 1996). The models are shown in order according to the number of parameters ($K$); however, this is only for convenience.

| Model | $K$ | $\log \mathcal{L}$ | AICc | $\Delta_i$ | $w_i$ | adj $R^2$ |
|---|---|---|---|---|---|---|
| 1. $ax^b$ | 3 | −110.82 | 227.64 | 813.12 | 0.0000 | 0.962 |
| 2. $a + b \log x$ | 3 | −42.78 | 91.56 | 677.04 | 0.0000 | 0.986 |
| 3. $a(x/(b + x))$ | 3 | −172.20 | 350.40 | 935.88 | 0.0000 | 0.903 |
| 4. $a(1 - e^{-bx})$ | 3 | −261.58 | 529.17 | 1114.65 | 0.0000 | 0.624 |
| 5. $a - bc^x$ | 4 | −107.76 | 223.53 | 809.01 | 0.0000 | 0.960 |
| 6. $(a + bx)/(1 + cx)$ | 4 | −24.76 | 57.53 | 643.01 | 0.0000 | 0.989 |
| 7. $a(1 - e^{-bx})^c$ | 4 | 25.42 | −42.85 | 542.63 | 0.0000 | 0.995 |
| 8. $a(1 - [1 + (x/c)^d]^{-b})$ | 5 | 216.04 | −422.08 | 163.40 | 0.0000 | 0.999 |
| 9. $a[1 - e^{-(b(x-c)^d)}]$ | 5 | 297.74 | −585.48 | 0 | 1.0000 | 0.999 |

$K$ is the number of parameters in the regression model plus 1 for $\sigma^2$
Some values of the maximized log-likelihood are positive because some terms were constant across models and were omitted.
The first eight model probabilities are 0 to at least 34 decimal places.

and is very convincing evidence that even the second best model has no plausibility (note, too, $\Delta_8 = 163.4$). The evidence ratio for model 4 (where adj $R^2 = 0.624$) is $1.1 \times 10^{242}$.

Adjusted $R^2$ values are useful as a measure of the proportion of the variation "in common" but are not useful in model selection (McQuarrie and Tsai 1998). Information-theoretic approaches show convincingly that all of the models are incredibly poor, relative to model 9 (the best model). This is a case where there is essentially no model selection uncertainty; the actual K–L best model among those considered, beyond any doubt, is model 9. Of course, everything is conditional on the set of hypotheses and their models. It is possible that a tenth model might be better yet; this is why the science objective is to let these sets evolve as more is learned. On the other hand, if model 9 was never considered, then model 8 would look very good relative to the other models. There are several reasons why adjusted $R^2$ is poor in model selection; its usefulness should be restricted to description.

## 4.8   The Effect of a Flood on European Dippers

Marzolin's data on the European dipper (*Cinclus cinclus*) have become a classic for illustrating various analytical issues (see Lebreton et al. 1992). Rationale for the hypotheses was outlined in Sect. 2.7 and this would be a point where this introductory material should be reread.

The problem focused on two models, one model without a flood effect on survival (model $\{\phi, p\}$) and another with a flood effect ($\{\phi_n, \phi_f, p\}$), where $\phi$ is the apparent annual probability of survival, $p$ is the annual probability of capture, and the subscripts denote a normal year ($n$) or a flood year ($f$). Under an information-theoretic approach, the results can be given very clearly in terms of the model probabilities (the $w_i$):

| Model | K | AICc | $\Delta_i$ | Probability, $w_i$ |
|---|---|---|---|---|
| $\{\phi,p\}$ | 2 | 670.866 | 4.706 | 0.0868 |
| $\{\phi_n,\phi_f,p\}$ | 3 | 666.160 | 0.000 | 0.9132 |

The table above provides the quantitative evidence and the evidence favors the hypothesis that annual survival decreased in the two years when a flood occurred. The evidence ratio is 10.5 = 0.9132/0.0868 indicating that the flood-affect hypothesis had about ten times more empirical support than the hypothesis that survival did not change during the years of the flood. [*Note*: I am careful to avoid any notion of causation here; however, the result can be said to be confirmatory because of the *a priori* hypotheses.]

Looking deeper, the MLEs for $\phi$ and measures of precision for the two models are:

| Model | | MLE | 95% confidence interval |
|---|---|---|---|
| $\{\phi,p\}$ | | 0.5602 | 0.5105–0.6888 |
| $\{\phi_n,\phi_f,p\}$ | ($n$) | 0.6071 | 0.5451–0.6658 |
| | ($f$) | 0.4688 | 0.3858–0.5537 |

The MLEs of the capture probability for the models were 0.9026 vs. 0.8998, respectively; these are virtually identical. The estimated "effect size" for survival was 0.1383 (=0.6071 – 0.4688), s.e. = 0.0532, and 95% confidence interval of (0.0340, 0.2425). The "evidence" in this case includes the model probabilities, an evidence ratio, estimates of effect size, and measures of precision. These entities allow a qualification of the hard, quantitative evidence.

In Sect. 2.7, we thought deeper about the issues and considered the following, alternative hypotheses:

- Was there a survival effect just the first year of the flood $\{\phi_{f1},\phi_{nf},p\}$?
- Or just the second year of the flood $\{\phi_{f2},\phi_{nf},p\}$?
- Or was the recapture probability ($p$) also effected by the flood $\{\phi_f,\phi_{nf},p_f,p_{nf}\}$?
- Or even $\{\phi,p_f,p_{nf}\}$ where survival was not impacted, but the recapture probabilities were?

We could now address these questions as if they were *a priori*; they should have been as they are questions begging to be asked. Instead, for example, we will assume the candidate set was not well thought out and, these four new hypotheses arose *post hoc* (a not unusual situation). Thus, we will admit the

TABLE 4.2. Model selection results for the European dipper data.

| Model | K | log $\mathcal{L}$ | AICc | $\Delta i$ | wi | Rank | "Tickets" |
|---|---|---|---|---|---|---|---|
| $\{\phi_n, \phi_f, p\}$ | 3 | **−330.051** | 666.160 | 0.000 | **0.559** | 1 | 1000[a] |
| $\{\phi_f, \phi_{nf}, p_f, p_{nf}\}$ | 4 | **−330.030** | 668.156 | 1.996 | **0.206** | 2 | 369 |
| $\{\phi_{f1}, \phi_{nf}, p\}$ | 3 | −331.839 | 669.735 | 3.575 | 0.094 | 3 | 167 |
| $\{\phi_{f2}, \phi_{nf}, p\}$ | 3 | −332.141 | 670.338 | 4.178 | 0.069 | 4 | 124 |
| $\{\phi, p\}$ | 2 | −333.419 | 670.866 | 4.706 | 0.053 | 5 | 95 |
| $\{\phi, p_f, p_{nf}\}$ | 3 | −333.412 | 672.881 | 6.721 | 0.019 | 6 | 35 |

[a]This column is just $1,000 \times \exp(-(1/2)\Delta_i)$ to illustrate the evidence for each of the six hypotheses in terms of how many raffle tickets each had. This is just a useful way to comprehend the evidence. Ticket numbers help in understanding; I am not proposing these be placed in publications

*post hoc* nature of the issue and examine a set of six models (the two original *a priori* models and the four new *post hoc* models). We will be prepared to tell our reader what we did and promise to treat these *post hoc* results as more tentative and speculative. The results are summarized in rank order in Table 4.2.

Note that the evidence ratio between model $\{\phi, p\}$ and model $\{\phi_n, \phi_f, p\}$ did not change ($0.559/0.053 = \exp(4.706/2) = 10.5$) as the four new models were added; however, the model probabilities ($w_i$) related to the candidate set change as models are added or deleted. Model probabilities are conditional on the candidate set.

The key question deals with a possible flood affect on the capture probabilities ($p_i$). The evidence ratio for $\{\phi_n, \phi_f, p\}$ vs. $\{\phi_n, \phi_f, p_f, p_{nf}\} = 0.559/0.206 = 2.7$. This evidence might be judged as weak; nonetheless, it does not support the notion that the capture probabilities varied with the year of the flood. Another piece of evidence lies in the MLEs and their profile likelihood intervals (Appendix A) for model $\{\phi_n, \phi_f, p_f, p_{nf}\}$:

| Parameter | Estimate | 95% profile interval |
|---|---|---|
| $p_{nf}$ | 0.904 | (0.836, 0.952) |
| $p_f$ | 0.893 | (0.683, 0.992) |

Overall, most would judge the evidence to be largely lacking; survival seems to have been influenced by the flood, but not the capture probabilities. It may also be interesting to note that the two models with no flood effect on survival were ranked last. If a much larger data set had been available, perhaps other effects would have been uncovered (i.e., the concept of tapering effect size, Sect. 2.3.5).

More careful examination of the table suggests another example of the "pretending variable" problem with model $\{\phi_f, \phi_{nf}, p_f, p_{nf}\}$. The addition of one additional parameter did not improve the fit as the log-likelihood value changed very little (−330.051 vs. −330.030) and the model was penalized by only $\Delta = 2$ units. This is further evidence that the flood had little effect on capture probabilities.

The data on dippers were taken by gender, thus science hypotheses could have related gender to the flood. For example, are females more likely to have lowered survival in flood years as perhaps they are tending the nest close to the water's edge? Perhaps the capture probability is higher for males as they forage over a wider area and more vulnerable to being caught. These and other questions are the type of thinking and hypothesizing that Chamberlin would have wanted – especially if it were done *a priori* to data analysis.

Although the dipper data and the science questions are fairly simple, they illustrate several key points. First, we wish to make an *inductive inference* from the sample data to the population of dippers along the streams in the years sampled. Second, that inference is *model based*. Third, the model probabilities and evidence ratios admit a high degree of rigor in the inferences. This approach refines Platt's (1964) concept of strong inference.

The data here are a series of 0s and 1s indicating the capture history of each dipper; simple plots of these histories would reveal nothing of interest. That is, the data are capture histories for each bird:

$$
\begin{array}{l}
11001101 \\
10001101 \\
00100001 \\
00000010
\end{array}
$$

for four birds in this example (see Appendix B). This is a clear case where inference must be model based as simple graphics would not be informative. Fourth, the models are products of multinomial distributions, and MLEs and their covariance matrix are derived from statistical theory. Fifth, the initial investigation was *confirmatory* and this allowed a more directed analysis and result. I chose to assume that the four additional hypotheses were *post hoc* although, ideally, they too would have been the result of *a priori* thinking and hypothesizing. Finally, perhaps no hypotheses (above) would warrant dismissal; no hypothesis seems to be virtually without support. Compare this situation with that from the disease transmission in ferrets where 2–3 of the hypotheses could easily be dismissed. Thus the model set for dippers might evolve by refinement of the existing models (not likely) or by further hypothesizing, but not by dropping some hypotheses and their models. Data on this population have been collected for several additional years and those data could be subjected to the same six hypotheses and analysis methods; at that time the candidate set might begin to evolve more rapidly.

## 4.9    More About Evidence and Inference

I find it useful to think about evidence about a parameter as the maximum likelihood estimate $\hat{\theta}$ and its profile likelihood interval (a type of confidence interval, see Appendix A). Both of these evidential quantities are dependent on a model. Traditionally, the model was assumed to be *given*. Now we have rigorous methods to select the model from the data. In Chap. 5, we will see that estimates of model parameters can be made from all the models in the set (multimodel inference) and this often has distinct advantages.

Less well known, but equally important, are the types of evidences about alternative science hypotheses.

---

**Types of Evidence**

There are three main kinds of evidences, in addition to simple ranking, concerning the evaluation of alternative science hypotheses:

1. Model probabilities, the probability that model $i$ is, in fact, the K–L best model. These are denoted as $w_i$ and they are also formal Bayesian posterior model probabilities.
2. The (relative) likelihood of model $i$, given the data. Such likelihoods are denoted as $\mathcal{L}(g_i|\text{data})$. Likelihoods are always relative to something else of interest; e.g., a likelihood of 0.31 means nothing by itself.
3. Evidence ratios provide the empirical evidence (or support) of hypothesis $i$ vs. $j$, $E_{ij} = w_i/w_j$; simply the ratio of model likelihoods or model probabilities for any two models $i$ and $j$.

---

The evidence ratio relates to any models $i$ and $j$, regardless of other models in or out of the set. Model probabilities depend on exactly $R$ hypotheses in the set; they are conditional on the set being fixed. Of course, all the three quantities stem from the differences, $\Delta_i$. It is the $\Delta_i$ that are on a scale of information and are the basis for the other measures of evidence.

A simple ranking of alternative science hypotheses is a form of evidence – ranking is based on the data and stems from "the information lost when using a hypothesis or model to approximate full reality." We want models that keep information loss to a minimum, as seen in Chap. 3.

None of these types of evidences are meant to be used in a dichotomous yes/no fashion. These are ways to quantify the evidence; this is where the science stops. From there, one can qualify the evidence to aid in understanding and interpretation. One should avoid thinking that, for example, "$\Delta_4$ is greater than 10, therefore it is unimportant or implausible" (or worse yet, "not significant"). There are always value judgments in interpreting evidence; often, virtually every objective person might arrive at the same value judgment, whereas in other cases, considered opinion (judgment) will vary substantially.

An important component of the analysis of data remains model assessment of the global model. Here, $R^2$, goodness-of-fit evaluations, and analysis of residuals have a role in assuring that some of the models are useful. When this is in doubt, it is sometimes useful to include a model with little or no structure and be sure that it is relatively implausible compared to, say, the global model.

## 4.10   Summary

Going back to 1890, Chamberlin asked, "What is the measure of probability on one side or the other?" Although it took nearly 100 years to develop methods to provide such probabilities, they are very useful in the evaluation of alternative science hypotheses. When data are entered into a proper likelihood and $-\log \mathcal{L}$ is computed, the units are "information," regardless of the original measurement units associated with the data.

I suspect Chamberlin might have been fairly content (in 1890) with a way to merely rank alternative hypotheses, based on the empirical data. Ranking is so interesting and compelling. Still, I find the concept of evidence ratios to be very central in evaluating the "strength of evidence" for alternative science hypotheses. This preference is not intended to downplay the model probabilities and Bayesians would certainly favor these in their framework. To me, the evidence ratios are so revealing and so comparative; however, I use the other quantities also.

It is conceptually important to recognize that the $\Delta_i$ define a narrow window of values likely to be judged as plausible. This window might reasonably be defined as 0 to 8–13 or so. This window exists regardless of the scale of measurement (e.g., millimeters or kilometers), the type of response variable (e.g., continuous, binomial), dimensionality of the parameter vector (3–8 or maybe 20–100), nested or nonnested models, and number of variables at hand. [This window assumes independence of the data (outcomes) – but see Sect. 6.2. Time series and spatial models provide a proper treatment for these dependent data.]

Every new method described in this chapter is simple to compute and understand; they also seem compelling. It is very important that scientists understand their results at a deep level. Biologists working in a team situation may often find that others on the team "did the analysis" and they have little or no idea as to what was done and for what reason. These are methods that should be relatively easy to comprehend and this is central to good science. Good science strategy tries to push the information gained to let the set evolve. This ever-changing set is the key to rapid advancement in knowledge and understanding.

Perhaps the biggest drawback to these approaches, as with all approaches, is the challenge to carefully define and model the hypotheses. If a model poorly reflects the science hypothesis, then everything is compromised. This is a continual challenge; I think statistics courses for both majors and nonmajors could better focus on modeling, rather than the current emphasis on null hypothesis testing.

Still the focus, from a science standpoint, must be on the alternative hypotheses. This focus is so central but so easily distracted. Investigators running "all possible models" via some powerful computer software have missed the entire point of the study. Models should arise to represent carefully derived science hypotheses; they should not arise just because the software makes them pos-

sible. Finally, it should be noted that the information-theoretic approach uni-
fies parameter estimation and the selection of a parsimonious approximating
model; both can be viewed as optimization problems.

## 4.11 Remarks

The idea of the likelihood of the model, given the data, was suggested
many years ago by Akaike (e.g., Akaike 1978b, 1979, 1980, 1981b; also
see Bozdogan 1987; Kishino 1991).

Royall's (1997) book focused on the Law of Likelihood and likelihood
ratios as evidence for one hypothesis over another. His treatment was useful
in cases for simple models and where the number of parameters in the two
models is the same, or perhaps differ by 1. This short book offers many valuable insights and is easy to follow, but is not a book for beginners.

It is helpful to recall that we are not just trying to fit the data; instead, we
are trying to recover the information in the data and allowing robust prediction
by using a model. It is model based inference and that inference comes from
a *fitted* model (i.e., the parameters have been estimated from the data, given
that model). These realities are important while trying to avoid the notion that
a model is true and fully represents reality.

AICc formulates the problem explicitly as a problem of *approximation*
of reality. Real data do not come from models. We cannot hope to find full
reality using models and finite data. As a crutch, we can think of full reality
as infinite dimensional; however, full reality is unlikely to be parameterized.
Parameters are a construct useful in many science contexts, but many parts of
full reality are not even parameterized. A "good" model successfully separates information from "noise" or noninformation in the data, but never fully
represents truth.

If a person insists that they have credible prior beliefs about each model
being the actual K–L best model, then these beliefs can be formally specified
as a set of discrete prior probabilities (in a Bayesian sense) on models, $\zeta_i$.
It seems unlikely that we would have a prior belief in this case as this would
entail a belief about approximations to full reality as well as the expected
parsimonious trade-offs in model fitting and how this varies by sample size.
The $\zeta_i$ must be the prior probabilities as to which model, *when fit to the data*
($\theta$ is estimated), is best for representing the (finite) information in the data.
If one had a set of priors on models ($\zeta_i$), then there is a simple way to generalize
the model probabilities to reflect this:

$$w_i = \frac{\exp\left(-\frac{1}{2}\Delta_i\right)\zeta_i}{\sum_{r=1}^{R}\exp\left(-\frac{1}{2}\Delta_r\right)\zeta_r}.$$

This is not a Bayesian approach and its properties are unknown (maybe unknowable?). Neither Burnham nor I would use this approach, but it does exist if people can honestly think they have informed beliefs about the model priors $\zeta_i$. More detail on this approach is given by Burnham and Anderson (2002:76–77).

It is clear now why Akaike considered the information-theoretic methods to be extensions of Fisher's likelihood theory. People believing in and using these methods might be classed as "likelihoodists" to distinguish themselves from traditional "frequentists" with their emphasis on null hypothesis testing and notions of confidence intervals and "Bayesians" with their priors on parameters and priors on models and inference being based on the posterior distribution.

Classifications such as this are sometimes helpful but perhaps the best philosophy is to use the best tool for a specific science problem. I do not hide the fact that I think the best, and most practical, approach is most often the objective one based on K–L information and entropy. I do not support the subjective aspects of the Bayesian methods in science but believe the objective Bayesian approach is well suited for the large class of random effects models (some additional comments are offered in Appendix D).

One often hears of "traditional" frequentist statistics – meaning null hypothesis testing approaches and frequentist confidence intervals, etc. It should be noted that although such approaches date back to the early parts of the twentieth century, the Bayesian approach goes back quite further, to the mid-1700s.

I do not favor the word "testing" regarding these new approaches, as it seems confusing and might imply testing null hypotheses. "Evaluate" seems a better choice of words. The word "significant" is best avoided in all respects.

Fisher tried to avoid a cutoff value to interpret his $P$-values. He once chose $\alpha = 0.05$ saying that a one in 20 chance might be something to suggest interest. Neyman and Pearson's approach called for $\alpha$ to be set in advance and was used in decisions about "significance" and power and so on. People have a natural tendency to simplify, but the reliance on (always arbitrary) cutoff points is to be discouraged in the empirical sciences.

The so-called Akaike weights are the basis for approaches to making inference from all the models in the set (Chap. 5), but they are interestingly more. Ken Burnham found that if one takes the Bayesian approach that Schwarz (1978) took and uses "savvy priors" on models instead of vague priors on models, AIC drops out (instead of BIC)! There is more to this at a technical level, but it is in this sense that it can be shown that AIC can be derived from Bayesian roots (Burnham and Anderson 2004). Thus, the $w_i$ are properly termed Bayesian posterior model probabilities. I usually prefer to call them just "model probabilities."

Bayesians are struggling with the issue of assigning prior probabilities on models and how to make these "innocent" or vague. It seems reasonable to investigate further the general notion of savvy or K–L priors on models in a Bayesian framework (see Burnham and Anderson 2004 for a more technical treatment of this issue).

The "theory of the theory" can get very deep in model selection (e.g., Linhart and Zucchini 1986; van der Linde 2004, and especially Massart 2007). Also see Vol. 50 of the *Journal of Mathematical Psychology*, edited by Wagenmakers and Waldorp, for some current research results.

## 4.12    Exercises

1. Reread the information in Sect. 2.4.1 on beak lengths in Darwin's finches. There were seven science hypotheses concerning possible evolutionary changes in beak lengths and these were represented by seven models. Your research team has collected data on beak lengths over many years and it is time for an analysis. Your technician has studied each of the models and has obtained the MLEs of the model parameters and the value of the maximized log-likelihood function. These are summarized below:

| Model $i$ | log £ | $K$ | AICc | $\Delta_i$ | $w_i$ |
|---|---|---|---|---|---|
| 1 | −66.21 | 2 | | | |
| 2 | −57.77 | 5 | | | |
| 3 | −59.43 | 6 | | | |
| 4 | −60.98 | 6 | | | |
| 5 | −49.47 | 6 | | | |
| 6 | −49.47 | 7 | | | |
| 7 | −49.46 | 8 | | | |

You are asked to complete the computations for the remaining columns and use the results as evidence in addressing the following questions:

a. What hypothesis is best supported? Why do you say it is the "best"?
b. Do these data provide evidence for two phenotypes? Why do you say this?
c. Is the evidence for two phenotypes strong? Weak? Be specific.
d. What is the supporting evidence for the covariates? Which one? Both? Why?
e. What is the evidence for the interaction term in model 7?
f. What further, *post hoc*, analyses might be considered? How would future research be molded by these results?

2. Some results concerning the affect of a flood on dippers were given in Sect. 4.8. The two models were nested and a likelihood ratio test was made: model $\{\phi,p\}$ vs. model $\{\phi_n,\phi_p,p\}$. The test statistic was 6.735 on 1 degree of freedom, giving a *P*-value of 0.0095. This is markedly different (by an order of magnitude!) from the probability of model $\{\phi,p\}$, given the data (0.0868). Explain this issue in some detail.

3. The nightly weather forecast uses the words probability of rain and likelihood of rain as synonyms. Detail the technical differences in these terms.

4. You are given the result, "the probability of hypothesis $H_4$, represented by its model $g_4$ is 0.53." This result might be called "conditional," but how? Explain. How is this different from an evidence ratio?

5. Can you intuit why adjusted $R^2$ leads to overfitted models? (advanced question)

6. Can you prove to yourself that the differencing leading to the $\Delta_i$ values removes the constant (across models) term that was omitted in the heuristic derivation from K–L information to AIC?

7. You have a colleague in Spain collaborating with you on an interesting science problem. You have worked together and have settled on six hypotheses, have derived six models to carefully reflect these hypotheses, and have fitted the data to the models. Thus, you have the MLEs, the covariance matrix, and other associated quantities. Everything seems to be in order. She has taken the lead in the analysis and provides the following AICc values:

$$H_1 \; 3,211 \qquad\qquad H_4 \; 14,712$$
$$H_2 \; 3,230 \qquad\qquad H_5 \; 7,202$$
$$H_3 \; 3,234 \qquad\qquad H_6 \; 5,699$$

She tentatively suggests that $H_1$ is the best, $H_2$ and $H_3$ are very close, but that $H_4$–$H_6$ are very poor (implausible, actually). Compose your e-mail comments on her findings so far. Form your response in terms of strength of evidence and ways to provide this.

8. While the mapping from the residual sum of squares (RSS) to the maximized $\log(L)$ is simple, many people stumble in trying to use the information-theoretic approaches in a simple "$t$-test" or ANOVA setting. First, consider a paired design of sample size $n$. (a) Lay out the models for the null and alternative hypotheses and display the computations for the RSS. (b) Lay out the same items for the unpaired design. Then, discuss how one would proceed to compute AICc, $\Delta_i$, Prob$\{H_0|data\}$, Prob$\{H_A|data\}$ and an evidence ratio. Finally, compare the advantages and differences between the usual $t$-statistics, the $P$-value, and rulings of statistical "significance" vs. model probabilities and evidence ratios.

# 5

# Multimodel Inference

Hirotugu Akaike was born in 1927 in Fujinomiya-shi, Shizuoka-jen in Japan. He received B.S. and D.S. degrees in mathematics from the University of Tokyo in 1952 and 1961, respectively. He worked at the Institute of Statistical Mathematics for over 30 years, becoming its Director General in 1982. He has received many awards, prizes, and honors for his work in theoretical and applied statistics (de Leeuw 1992; Parzen 1994). This list includes the Asahi Prize, the Japanese Medal with Purple Ribbon, the Japan Statistical Society Award, and the 2006 Kyoto Prize. The three volume set entitled *"Proceedings of the First US/Japan Conference on the Frontiers of Statistical Modeling: An Informational Approach"* (Bozdogan 1994) was to commemorate Professor Hirotugu Akaike's 65th birthday. He is currently a Professor Emeritus at the Institute, a position he has held since 1994 and he received the Kyoto Prize in Basic Science in March, 2007.

For many years it seemed logical to somehow select the best model from an *a priori* set (but many people ran "all possible models") and make inductive inferences from that best model. This approach has been the goal, for example, in regression analysis using AIC, Mallows' (1973) $C_p$ or step-up, step-down, or stepwise methods. Making inferences from the estimated best model seems logical and has served science for the past 50 years.

It turns out that a better approach is to make inferences from all the science hypotheses and their associated models in an *a priori* set. Upon deeper reflection, perhaps it is not surprising that one might want to make inferences from all the models in the set, but it is very surprising that this expanded approach is so conceptually simple and easy to compute. Approaches to making formal inferences from all (or, at least, many) models is termed *multimodel inference*.

Multimodel inference is a fairly new issue in the statistical sciences and one can expect further advances in the coming years. At this time, there are four aspects to multimodel inference: model averaging, unconditional variances, gauging the relative importance of variables, and confidence sets on models.

## 5.1   Model Averaging

There are theoretical reasons to consider multimodel inference in general and model averaging in particular. I will not review these technicalities and instead offer some conceptual insights on this interesting and effective approach.

First, it becomes clear from the model probabilities ($w_i$) and the model likelihoods ($\mathcal{L}(g_i|\text{data})$) that there is relevant *information* in models ranked below the (estimated) best model. Although there are exceptions (e.g., Flather's data in Sect. 4.7), there is often a substantial amount of information in models ranked second, third, etc. (e.g., the dipper data in Table 4.2). If there is information in the model ranked second, why not attempt to use it in making inferences? Multimodel inference captures this information by using all, or several of, the models in the set.

Second, most models in the life sciences are far from full reality. With fairly adequate sample sizes, we might hope to find first- and perhaps second-order effects and maybe low-order interactions. Our models are often only crude approximations: unable to get at the countless smaller effects, nonlinearities, measured covariates, and higher order interactions. For example, even the best of the Flather models with an $R^2$ value of 0.999 is hardly close to full reality for the system he studied (i.e., dozens of species of birds, across several states and years, with data taken from a wide variety of observers). So, we might ask why we want to base the entire inference on the (estimated) best model when there is usually uncertainty in the selection as to the "best" model. Instead, perhaps inference should be based on a "cloud" of models, weighted in some way such that better models have more influence than models that are relatively poor. Here, rough classification of "good" and "poor" models is empirically based using model probabilities and model likelihoods. These lines of reasoning lead to what has been termed *model averaging*. Although the general notion has been in the statistical literature for many years, Bayesians have championed this approach in the past 10–15 years; however, their approach is computationally quite different.

Third, *ad hoc* rules such as "$\Delta_i$ greater than 2" become obsolete as all the models are used for inference. A model-averaged regression equation attempts to display the smaller effects and one can judge their usefulness against the science question of interest.

## 5.1.1    Model Averaging for Prediction

The best way to understand model averaging is in prediction. Consider the case where one has four well-defined science hypotheses $(H_1,\dots, H_4)$ and these are represented by four regression models $(g_1,\dots, g_4)$. The data are analyzed using standard least squares methods to obtain the parameter estimates $(\hat{\beta}, \hat{\sigma}^2)$ and their covariance matrix $(\Sigma)$. For a given set of values of the predictor values (say, $x_1 = 4.1, x_2 = 3.3, x_3 = 0.87$, and $x_4 = -4.5$), one can use any of the models to make a prediction $(\hat{Y})$ of the response variable $Y$, for example,

$$\hat{Y} = \hat{\beta}_0 + \hat{\beta}_1(x_1) + \hat{\beta}_3(x_3) + \hat{\beta}_4(x_4)$$

specifically,

$$\hat{Y} = \hat{\beta}_0 + \hat{\beta}_1(4.1) + \hat{\beta}_3(0.87) + \hat{\beta}_4(-4.5)$$

using the values given above (note that this model does not include $x_2$). Of course, the MLEs of the $\beta_j$ would have to be available before this model could be used for prediction of the response variable $(Y)$.

Assume that this is the best model and has a model probability of 0.66. The second-best model excludes $x_1$ and is

$$\hat{Y} = \hat{\beta}_0 + \hat{\beta}_3(x_3) + \hat{\beta}_4(x_4) \quad \text{or} \quad = \hat{\beta}_0 + \hat{\beta}_3(0.87) + \hat{\beta}_4(-4.5)$$

and has a model probability of 0.27. The $\beta_i$ parameters differ somewhat by model; that is, $\beta_2$ in the models above are different. I trust that notation to make this difference explicit is not needed as it would clutter the simple issue (i.e., the numerical value of $\hat{\beta}_3$ in the best model is almost surely different from the value of $\hat{\beta}_3$ in the second model). These and the other two models are summarized by rank as

| Model $i$ | $\hat{Y}_i$ | Model probability $w_i$ |
| --- | --- | --- |
| 1 | 67.0 | 0.660 |
| 2 | 51.7 | 0.270 |
| 3 | 54.5 | 0.070 |
| 4 | 71.1 | <0.001 |

The predicted values can vary substantially by model; the $\beta_j$ typically vary less from model to model. Prediction based on the best model might be risky as it has only two-thirds of the model probability and prediction from this model is somewhat higher than the other two models having the remaining one-third of the weight. Thus, the urge to make inference concerning predicted values using all four models (multimodel inference).

In this case, model-averaged prediction is a simple sum of the model probability for model $i$ $(w_i)$ times the predicted value for model $i$ $(Y_i)$

$$\hat{\bar{Y}} = \sum_{i=1}^{R} w_i \hat{Y}_i,$$

where $\hat{\bar{Y}}_i$ denotes a model-averaged prediction of the response variable $Y$ and $R = 4$ in this example. This is merely

$$0.660 \times 67.0 + 0.270 \times 51.7 + 0.070 \times 54.5 + 0.000 \times 71.1 = 61.994$$

or 62 for practical purposes. The model-averaged prediction downweights the prediction from the (estimated) best model for the fact that the other two models provide lower predicted values and they carry one-third of the weight. The fourth model has essentially no weight and does not contribute to the model-averaged prediction. Model-averaging prediction is trivial to compute; it is a simple weighted average. In the past, people did not know what to use for the weights and an unweighted average made little sense. The proper weights are just the model probabilities and $\hat{\bar{Y}}$ is fairly robust to slight differences in the weights. For example, bootstrapped weights (see Burnham and Anderson 2002:90–94) or weights from BIC (Appendix E) are different somewhat from those defined here; however, these differences often make relatively little change in the model-averaged estimates).

---

### Model Averaging Predictions

Model averaging for prediction is merely a weighted average of the predictions ($\hat{Y}$) from each of the R models.

$$\hat{\bar{Y}} = \sum_{i=1}^{R} w_i \hat{Y}_i,$$

where the subscript $i$ is over models. The weights are the model probabilities ($w_i$) and the predicted values ($Y_i$) from each model are given a particular setting of the predictor variables.

---

Typically, all the models in the set are used in such averaging. If one has good reason to delete one or more models, then the model weights must be renormalized such that they sum to 1.

I have been asked if some of the poorest models have "bad" information and, thus, should be excluded in the averaging. Lacking absolute proof, I have recommended against this. For one thing, if a model is that poor, it receives virtually no weight. I generally prefer using all the models in the *a priori* set in model averaging as this keeps unwanted subjectivity out of the process. Prediction is the ideal way to understand model averaging because each model can always be made to make an estimate of the response variable, given settings of the predictor variables.

### 5.1.2   Model Averaging Parameter Estimates Across Models

There are many cases where one or more model parameters are of special interest and each of the models has the parameter(s). In such cases, model averaging may provide a robust estimate.

---

**Model Averaging Parameters within Models**

An obvious approach to gain robust estimates of model parameters is to model average the estimates ($\hat{\theta}$) from each of the models $i$, where i = 1, 2, ...., R,

$$\hat{\bar{\theta}} = \sum_{i=1}^{R} w_i \hat{\theta}_i,$$

where $\hat{\bar{\theta}}$ is the model averaged estimate of $\theta$ ($\theta$ is used to denote some generic parameter of interest)

---

I will use the data on European dippers from Sect. 4.8 to illustrate averaging of estimates of model parameters. I found the MLEs and model probabilities for the two models to be the following:

| Model | MLE | Model probabilities $w_i$ |
|---|---|---|
| $\{\phi,p\}$ | 0.5602 | 0.08681 |
| $\{\phi_n,\phi_f,p\}$ $(n)$ | 0.6071 | 0.91319 |
| $(f)$ | 0.4688 | |

Model averaging for the survival probability in normal years is just

$$\hat{\bar{\phi}} = \sum_{i=1}^{2} w_i \hat{\phi}_i,$$

or

$$0.08681 \times 0.5602 + 0.91319 \times 0.6071 = 0.6030,$$

and the model-averaged estimate of the survival probability in flood years is

$$0.08681 \times 0.5602 + 0.91319 \times 0.4688 = 0.4767.$$

Note that the MLE for both flood and nonflood years in model $\{\phi,p\}$ is merely $\phi = 0.5602$, because this model does not recognize a flood year as being different from any other year. In this particular example, the estimates changed little because the best model had nearly all of the weight (91%). If the best model has a high weight (e.g., 0.90 or 0.95), then model averaging will make little difference as the other models contribute little to the average because they have virtually no weight (an exception would be the case where the estimates for models with little weight are very different than the other estimates). In other words, there is relatively little model selection uncertainty in this case and the model-averaged estimate is similar to the numerical value from the best model.

There are many cases where the parameter of interest does not appear in all the models and this important case is more complicated and is the subject of Sect. 6.2. Although not as easy to conceptualize or compute, there are ways to cope with the difficult issue of model selection bias (Sect. 6.2.2).

The investigator must decide if model averaging makes sense in a particular application. In general, I highly recommend model averaging in cases where there is special interest in a parameter or subset of parameters. It should be clear that the parameter being considered from model averaging must mean the same thing in each model. Thus, if there is special interest in a disease transmission probability $\lambda$, then this parameter must be consistent in its meaning across models.

The models used by Flather have parameters $a$, $b$, $c$, and $d$ but these mean very different things from one model to another. For example, consider three of his models:

$$E(Y) = ax^b,$$

$$E(Y) = a + \log(x), \text{ and}$$

$$E(Y) = a\left(1 - [1 + (x/c)^d]^{-b}\right).$$

The parameter $a$, for instance, has quite different roles in these three models and it makes no sense to try to average the estimates $\hat{a}$ across models.

## 5.2   Unconditional Variances

Sampling variances are a measure of precision or repeatability of an estimator, *given* the model and the sample size. In the model above

$$E(\hat{Y}) = \hat{\beta}_0 + \hat{\beta}_3(x_3) + \hat{\beta}_4(x_4).$$

The variance of the estimator $\hat{\beta}_3$ is said to be "conditional" on the model. In other words, *given* this model, one can obtain a numerical value for $\text{var}(\hat{\beta}_3)$ using least squares or maximum likelihood theory. The fact is, another variance component should be included – a variance component due to *model selection uncertainty*. In the real world, we are not given *the* model and thus we cannot condition on the single best model resulting from data based selection. Instead, we use information-theoretic criteria or Mallows $Cp$ or *ad hoc* approaches such as stepwise regression to *select* a model from a set of models. Usually, there is uncertainty as to the best model selected in some manner. That is, if we had other replicate data sets of the same size and from the same process, we would often find a variety of models that are "best" across the replicate data sets. This is called *model selection uncertainty*." Model selection uncertainty arises from the fact that the data are used both to select a model and to estimate its parameters. Much has been written about problems that arise from this joint use of the data. The information-theoretic approaches have at least partially resolved these issues (the same can be said for several Bayesian approaches).

Thus, a proper measure of precision or repeatability of the estimator $\hat{\beta}_3$ must include both the usual sampling variability (i.e., given the model) and a measure of model selection uncertainty (i.e., the model to model variability in the estimates of $\hat{\beta}_3$). This issue has been known for many years; statisticians have noted that using just the sampling variance, given (i.e., "conditional on") the model represents a "quiet scandal." Ideally,

$$\mathrm{var}\left(\hat{\theta}\right) = \text{sampling variance given a model} +$$
$$\text{variation due to model selection uncertainty.}$$
$$= \mathrm{var}\left(\hat{\theta}_i | g_i\right) + \sum\left(\hat{\theta}_i - \hat{\bar{\theta}}\right)^2.$$

The last term captures the variation in the estimates of a particular parameter $\theta$ across models. If estimates of $\theta$ vary little from one model to another, this term is small and the main variance component is the usual one, $\mathrm{var}(\hat{\theta} \mid g)$, which is the variance of $\hat{\theta}$, conditional (given) on the model $g_i$. However, as is often the case, $\hat{\theta}$ varies substantially among models and if this term is omitted, the estimated variance is too small, the precision is overestimated, and the confidence intervals are too narrow.

Building on these heuristics, we can think of the above equation and take a weighted sum across models as

$$\mathbf{var}\left(\hat{\bar{\theta}}\right) = \sum_{i=1}^{R} w_i \left\{ \mathbf{var}\left(\hat{\theta}_i \mid g_i\right) + \left(\hat{\theta}_i - \hat{\bar{\theta}}\right)^2 \right\}$$

where the final term is the model-averaged estimator. This variance is called "unconditional" as it is not conditional on a *single* model; instead, it is conditional on the set of models considered (a weaker assumption). [Clearly, the word "unconditional" was poorly chosen; we just need to know what is meant by the term.] Note that model averaging arises as part of the theory for obtaining an estimator of the unconditional variance. The theory leading to this form has some Bayesian roots.

Note that the sum of the two variance components is weighted by the model probabilities. If the best model has, say, $w_{\text{best}} > 0.95$, then one might ignore the final variance component, $\left(\hat{\theta}_i - \hat{\bar{\theta}}\right)^2$, because model selection uncertainty is nil. In such cases, model averaging is not required and the usual conditional variance should suffice.

There are sometimes cases where the investigator wants to make inference about a parameter from only the best model; here an unconditional variance and associated confidence intervals could still be used with the expected advantages.

---

### Unconditional Variance Estimator

An estimator of the variance of parameter estimates that incorporates both sampling variance, given a model, and a variance component for model selection uncertainty is

$$\mathbf{var}\left(\hat{\bar{\theta}}\right) = \sum_{i=1}^{R} w_i \left\{ \mathbf{var}\left(\hat{\theta}_i \mid g_i\right) + \left(\hat{\theta}_i - \hat{\bar{\theta}}\right)^2 \right\},$$

where $\hat{\bar{\theta}}$ is the model averaged estimate, $w_i$ are the model probabilities, and $g_i$ is the $i$th model. This expression is also useful in a case where one is interested in a proper variance for $\hat{\theta}$ from only the best model (not the model-averaged estimator).

Of course $\mathrm{se}\left(\hat{\bar{\theta}}\right) = \sqrt{\mathrm{var}\left(\hat{\bar{\theta}}\right)}$, and if the sample size is large, a 95% confidence interval can be approximated as $\hat{\bar{\theta}} \pm 1.96 \times \mathrm{se}\,\hat{\bar{\theta}}$.

Use of this approach provides estimators with good achieved confidence interval coverage. I highly recommend this approach when data are used to both select a model and estimate its parameters (i.e., the usual case in applied data analysis).

### 5.2.1  Examples Using the Cement Hardening Data

We return to the cement hardening data to provide useful insights into model selection uncertainty for this small ($n = 13$) data set. Material presented in Sect. 4.5 indicated that only two models (models {12} and {12 1*2} had any discernible empirical support. This is almost too simple to illustrate the points I want to make clear; therefore, I will use an extended example using these data from Burnham and Anderson (2002:177–183).

Model averaging the $\beta_i$ parameters and obtaining the unconditional variances was done using, strictly for illustration, all possible subsets, thus there are $2^4-1=15$ models. Clearly, model {12} is indicated as the best; however, substantial model selection uncertainty is evident because that best model has a model probability of only 0.567 (Table 5.1).

TABLE 5.1. Model probabilities for 15 nested models of the cement hardening data.

| Model $i$ | $w_i$ | Model $i$ | $w_i$ |
|-----------|-------|-----------|-------|
| {12} | 0.5670 | {23} | 0.0000[a] |
| {124} | 0.1182 | {4} | 0.0000 |
| {123} | 0.1161 | {2} | 0.0000 |
| {14} | 0.1072 | {24} | 0.0000 |
| {134} | 0.0810 | {1} | 0.0000 |
| {234} | 0.0072 | {13} | 0.0000 |
| {1234} | 0.0029 | {3} | 0.0000 |
| {34} | 0.0004 | | |

[a]Values shown are zero to at least five decimal places.

TABLE 5.2. Quantities used to compute the unconditional variance of the predicted value for the cement hardening data.

| Model $i$ | $K$ | $\hat{Y}$ | $w_i$ | $\text{var}(\hat{Y} \mid g_i)$ | $(\hat{Y}_i - \hat{\bar{Y}})^2$ |
|-----------|-----|-----------|-------|-------------------------------|--------------------------------|
| {12} | 4 | 100.4 | 0.5670 | 0.536 | 1.503 |
| {124} | 5 | 102.2 | 0.1182 | 2.368 | 0.329 |
| {123} | 5 | 100.5 | 0.1161 | 0.503 | 1.268 |
| {14} | 4 | 105.2 | 0.1072 | 0.852 | 12.773 |
| {134} | 5 | 105.2 | 0.0810 | 0.643 | 12.773 |
| {234} | 5 | 111.9 | 0.0072 | 4.928 | 105.555 |
| {1234} | 6 | 101.6 | 0.0029 | 27.995 | 0.001 |
| {34} | 4 | 104.8 | 0.0004 | 1.971 | 10.074 |

The computation of unconditional estimates of precision for a predicted value is simple because every model $i$ can be made to provide a prediction ($\hat{Y}$). We consider prediction where the predictor values are set at $x_1 = 10, x_2 = 50,$ $x_3 = 10, x_4 = 20$. The prediction under each of the eight models is shown in Table 5.2 (the last seven models were dropped here as they have virtually no weight). Clearly, $\hat{Y}$ is high for model $\{234\}$, relative to the other models. The estimated standard error for model $\{1234\}$ is very high, as might be expected because the $X$ matrix is nearly singular. Both of these models have relatively little support, as reflected by the small model probabilities, and so the predicted values under these fitted models are of little credibility.

Table 5.2 predicts the response variable $Y$ for the cement hardening data. $\hat{Y}$ is a predicted expected response based on the fitted model (given the predictor values, $x_1 = 10, x_2 = 50, x_3 = 10,$ and $x_4 = 20$), conditional on the model. Measures of precision are given for $\hat{Y}$; $\bar{\hat{Y}}$ denotes a model-averaged predicted value, and $(\hat{Y}_i - \bar{\hat{Y}})^2$ is the model-to-model variance component when using model $i$ to estimate $Y$. For example, for model $\{12\}$ and $i = 1$, $(\hat{Y}_1 - \bar{\hat{Y}})^2$.

The remaining seven models had essentially no weight and are not shown.

The predicted value for the AICc-best model is 100.4 with an estimated conditional variance of 0.536. However, this measure of precision is an underestimate because the variance component due to model selection uncertainty has not been incorporated. Model averaging results in a predicted value of 101.6. The corresponding estimated unconditional standard error is 1.9. The unconditional standard error is substantially larger than the conditional standard error 0.73. Although it is not a virtue that the unconditional standard error is larger than has been used traditionally, it is a more honest reflection of the actual precision. If only a conditional standard error is used, then confidence intervals are too narrow and achieved coverage will often be substantially less than the nominal level (e.g., 95%).

Study of the final two columns in Table 5.2 shows that the variation in the model-specific predictions (i.e., $\hat{Y}$) from the weighted mean (i.e., $(\hat{Y}_i - \bar{\hat{Y}})^2$) is substantial relative to the estimated variation, conditional on the model (i.e., the $\text{var}(\hat{Y} \mid g_i)$). Models $\{124\}$ and $\{1234\}$ are exceptions in this case.

The investigator has the choice as to whether to use the predicted value from the AICc-selected model (100.4) or a model-averaged prediction (101.6). In this example, the differences in predicted values are small relative to the unconditional standard errors (1.9); thus, here the choice makes little difference. However, there is often considerable model uncertainty associated with real data and I would suggest the use of the model-averaged predictions. Thus, I would use 101.6 as the predicted value with an unconditional standard error of 1.9. If the best model was more strongly supported by the data (i.e., $w_i > 0.95$), then I might suggest the use of the prediction based on that (best) model (i.e., $\hat{Y} = 100.4$) and use the estimate of the unconditional standard error (1.9).

TABLE 5.3. Quantities needed to make multimodel inference concerning the parameter $\beta_1$ for the cement hardening data.

| Model $i$ | $\hat{\beta}_1$ | var($\hat{\beta}_1 \mid g_i$) | $w_i$ |
|---|---|---|---|
| {12} | 1.4683 | 0.01471 | 0.5670 |
| {124} | 1.4519 | 0.01369 | 0.1182 |
| {123} | 1.6959 | 0.04186 | 0.1161 |
| {14} | 1.4400 | 0.01915 | 0.1072 |
| {134} | 1.0519 | 0.05004 | 0.0811 |
| {1234} | 1.5511 | 0.55472 | 0.0029 |
| {1} | 1.8687 | 0.27710 | <0.0001 |
| {13} | 2.3125 | 0.92122 | <0.0001 |

Model {1} has only $x_1$ in it, allowing a "straight shot" at $\hat{\beta}_1$; however, it is a very poor model, relative to the others. When other variables are included in a model with $x_1$, the parameter estimates change due to the fact that the $x_1$ is correlated with the other predictor variables (see Table 5.3). This is frustrating as the biologist wants a simple answer to this simple question. So, what is a robust estimate of the regression slope on $x_1$? Model averaging is one approach to answer this question.

The computation of the model-averaged estimates of $\beta_1$ and its unconditional sampling variance is illustrated in Table 5.3 (recall that the variance is the square of the standard error).

The other seven models do not have $\beta_1$ in them and are not shown above.

Note the variation in the estimates of $\beta_1$ across models – this model-to-model variation is model selection uncertainty and needs to be reflected in estimates of precision. The model-averaged estimate,

$$\hat{\bar{\beta}}_1 = \sum_{i=1}^{R} w_i \hat{\beta}_{1i} = 1.4561,$$ while the estimate of $\beta_1$ from model {1} is 1.8687.

The unconditional variance of this model-averaged estimate of $\beta_1$ is obtained by using the formula above, expressed in terms of $\beta_1$ (instead of the generic parameter $\theta$),

$$\text{var}\left(\hat{\bar{\beta}}_1\right) = \sum_{i=1}^{R} w_i \left\{ \text{var}\left(\hat{\beta}_{1i} \mid g_i\right) + \left(\hat{\beta}_{1i} - \hat{\bar{\beta}}_1\right)^2 \right\},$$

where $i$ reflects the model and "1" is the parameter of interest, $\beta_1$. For example, the first term in the needed sum is

$$w_1 \times \left\{ \hat{\text{var}}\left(\beta_{11} \mid g_1\right) + \left(\hat{\beta}_{11} - \hat{\bar{\beta}}_1\right)^2 \right\}$$

$$0.5670 \times \{(0.0147) + (0.0122)^2\} = 0.00842.$$

Note that the final term is $1.4683 - 1.4561 = 0.0122$; then this is squared. Completing the rest of the calculations, we get var($\hat{\bar{\beta}}_1$) = 0.0308, or an estimated unconditional standard error on $\hat{\bar{\beta}}_1$ of 0.1755. This compares to the

TABLE 5.4.  Quantities needed to make multi-model inference concerning the parameter $\beta_2$ for the cement hardening data.

| Model | $\hat{\beta}_2$ | $se(\hat{\beta}_2 \mid g_i)$ | $w_i$ |
|---|---|---|---|
| {12} | 0.6623 | 0.0459 | 0.6988 |
| {124} | 0.4161 | 0.1856 | 0.1457 |
| {123} | 0.6569 | 0.0442 | 0.1431 |
| {1234} | 0.5102 | 0.7238 | 0.0035 |
| {234} | −0.9234 | 0.2619 | 0.0089 |
| {23} | 0.7313 | 0.1207 | <0.0001 |
| {24} | 0.3109 | 0.7486 | <0.0001 |
| {2} | 0.7891 | 0.1684 | <0.0001 |

conditional standard error of 0.1213, given the selected model. This difference is the "quiet scandal" because model selection uncertainty (a component of variance) had been omitted (ignored) in the conditional approach.

Model selection uncertainty is more clearly present when examining the model-to-model variation in the estimator of $\beta_2$ (Table 5.4). Note that the estimate of $\beta_2$ for model {234} is negative; this is due to the high correlations among the predictor variables.

The Akaike weights ($w_i$) for just the eight models above add to 0.8114. However, to compute results relevant to just these eight models, we must renormalize the relevant model probabilities to add to 1. Those renormalization probabilities are given in Table 5.4. The model-averaged estimator of $\beta_2$ is 0.6110 (compared to 0.6623 for the best model and 0.7891 for model {2}) and the unconditional estimated standard error of $\bar{\hat{\beta}}_2$ is 0.1206. The conditional standard error for the best model was 0.0459, while the unconditional standard error was 0.1206 – reflecting the high degree of model selection uncertainty. This estimate attempts to provide a robust estimate of the "slope" on the variable $x_2$ without regard to other variables in the model.

It is important to compute and use unconditional standard errors in inferences after data based model selection. Note also that confidence intervals on $\beta_1$ and $\beta_2$, using results from model {12}, should be constructed based on a $t$-statistic with 10 df ($t_{10,0.975} = 2.228$ for a two-sided 95% confidence interval). Such intervals here will still be bounded well away from 0; for example, the 95% interval for $\beta_2$ is 0.39–0.93.

## 5.2.2  Averaging Detection Probability Parameters in Occupancy Models

Ball et al. (2005) used occupancy models (MacKenzie et al. 2006) to evaluate a habitat model for the Palm Springs ground squirrel (*Spermophilis tereticaudus chlorus*) in the Coachella Valley, California, in 2002. I will use this paper to illustrate several things, in addition to model averaging and unconditional variances.

Ball et al. (2005) were interested in the evaluation of a habitat model developed by a Scientific Advisory Committee (SAC) as part of a multispecies Habitat Conservation Plan. Interest was focused on mesquite (*Prosopis glandulosa*), which was not common, thought to be excellent habitat, and in decline, while creosote (*Larrea tridentata*) was much more common but of somewhat questionable value compared to the squirrel.

Ball et al. (2005) developed 15 models based on squirrel occupancy ($\psi$) and squirrel detectability ($p$). They justified their choice of *a priori* hypotheses based largely on the literature, knowledge of the biology of the species, and on management concerns and this required considerable thought (Doherty, personal communication). A pilot study was conducted and these results used to improve survey design and data-gathering protocols. Four preserves were sampled using systematic samples with random starts. Over 1,900 points were sampled in the initial field session. Occupancy was modeled as a constant ($\cdot$) or as a function of individual vegetation (mesquite, creosote, or desert scrub) or substrate types (dune and hummock) or combinations. Detection probability was modeled as a function of these same variables and sampling time ($t$); sampling was done on three occasions ($t = 1, 2, 3$). Our entree into this issue will be Table 5.5 where 15 models are summarized. One thing to notice at first glance is that virtually 100% of the model probability is tied to just three models and the science hypotheses that they represent.

Estimates of the constant detection probability from the best model was 0.216 but varied from 0.156 to 0.256 for other models in the set. Model averaging was done and the results are shown in Table 5.6.

TABLE 5.5. Model selection statistics for 15 models of ground squirrel occupancy ($\psi$) and detection probability ($p$) from Ball et al. (2005). The models are shown by rank.

| | Hypothesis/model | log($L$) | AICc | $K$ | $\Delta_i$ | $w_i$ |
|---|---|---|---|---|---|---|
| 1 | $\psi_{mdh,cdh,sdh,ae} \cdot p_{\cdot}$ | −229.27 | 468.56 | 5 | 0.00 | 0.58 |
| 2 | $\psi_{mdh,cdh,sdh,ae} \cdot P_{mdh=cdh,ae}$ | −229.11 | 470.26 | 6 | 1.70 | 0.25 |
| 3 | $\psi_{mdh,cdh,sdh,ae} \cdot P_t$ | −228.49 | 471.05 | 7 | 2.48 | 0.17 |
| 4 | $\psi_{mdh=cdh,sdh,ae} \cdot p_{\cdot}$ | −235.32 | 478.66 | 4 | 10.09 | <0.01 |
| 5 | $\psi_{mdh=cdh,ae} \cdot p_{\cdot}$ | −236.97 | 479.96 | 3 | 11.40 | <0.01 |
| 6 | $\psi_{m,distom} \cdot p_{\cdot}$ | −236.00 | 480.01 | 4 | 11.45 | <0.01 |
| 7 | $\psi_{mdh=cdh,sdh,ae} \cdot P_{mdh=cdh,ae}$ | −235.31 | 480.65 | 5 | 12.09 | <0.01 |
| 8 | $\psi_{mdh=cdh,sdh,ae} \cdot P_t$ | −234.93 | 481.91 | 6 | 13.35 | <0.01 |
| 9 | $\psi_{mdh=cdh,sdh,ae} \cdot P_{mdh=cdh,ae}$ | −236.96 | 481.95 | 4 | 13.38 | <0.01 |
| 10 | $\psi_{distom} \cdot P_t$ | −235.25 | 482.54 | 6 | 13.98 | <0.01 |
| 11 | $\psi_{mdh=cdh,ae} \cdot P_t$ | −236.57 | 483.17 | 5 | 14.60 | <0.01 |
| 12 | $\psi_{\cdot} \cdot P_{mdh=cdh,ae}$ | −242.19 | 490.17 | 3 | 21.83 | <0.01 |
| 13 | $\psi_{\cdot} \cdot p_{\cdot}$ | −278.61 | 561.23 | 2 | 92.67 | <0.01 |
| 14 | $\psi_{\cdot} \cdot p_t$ | −278.36 | 564.74 | 4 | 96.18 | <0.01 |
| 15 | $\psi_{m,distom} \cdot P_{mdh=cdh,ae}$ | −279.66 | 569.34 | 5 | 100.78 | <0.01 |

*M* mesquite; *C* creosote; *D* dune; *H* hummock; *S* desert scrub (e.g., *Atriplex* spp.); *AE* all else; *DisToM* distance to mesquite.

TABLE 5.6. Summary of model averaging for detection probability for the ground squirrel data from Ball et al. (2005).

| | Estimated | | |
|---|---|---|---|
| Model number | Model probability | Detection probability | Standard error |
| 1 | 0.576 | 0.216 | 0.044 |
| 2 | 0.247 | 0.228 | 0.051 |
| 3 | 0.166 | 0.256 | 0.064 |
| 4 | 0.004 | 0.161 | 0.050 |
| 5 | 0.002 | 0.162 | 0.050 |
| 6 | 0.002 | 0.195 | 0.040 |
| 7 | 0.001 | 0.156 | 0.061 |
| 8 | 0.001 | 0.178 | 0.062 |
| 9 | 0.001 | 0.156 | 0.061 |
| 10 | 0.001 | 0.231 | 0.058 |
| 11 | <0.001 | 0.180 | 0.063 |
| 12[a] | <0.001 | 0.209 | 0.063 |
| Weighted average | | 0.225[b] | 0.049[c] |
| Unconditional standard error[d] | | | 0.052 |

[a] The remaining models had virtually no weight and are not shown. The results are shown to only three places.
[b] The weighted average was based on

$$\hat{\bar{p}} = \sum_{i=1}^{R} w_i \hat{p}_i,$$

[c] The first entry is a weighted average of the conditional standard errors, while the unconditional standard error includes a variance component for model selection uncertainty.
[d] The unconditional standard error was based on

$$\text{var}\left(\hat{\bar{p}}\right) = \sum_{i=1}^{R} w_i \left\{ \text{var}\left(\hat{p}_i \mid g_i\right) + \left(\hat{p}_i - \hat{\bar{p}}\right)^2 \right\}.$$

Approximately 11% of the variation stems from model selection uncertainty and is a small proportion in this example. Note that the weighted average of the conditional standard errors (0.049) is larger than the conditional standard error for the best model (0.044).

Careful examination of the log-likelihood values suggests that model 2 is a good model only because it has one additional parameter (thus a "penalty term" of approximately 2); however, the fit did not improve. That is, the log-likelihood value for the best model was −229.27, whereas this value for the second-best model was −229.11. This finding leads to the conclusion that the structure on the detection probability ($p_{\text{MDH=CDH,AE}}$) is without support. The two estimates of detection probability under the second-best model are similar (0.228 vs. 0.170) and their confidence intervals overlap entirely (0.144–0.342 vs. 0.059–0.403). This is an example of a "pretending variable" (see Sect. 3.6.8). One should check to be sure that there has been a change in the log-likelihood values to avoid the "pretending variable problem."

If model 2 is removed from the set for some *post hoc* reason, the model probabilities for the first- (1) and the second-best (formerly 3) models change to 0.764 and 0.221, respectively, from 0.576 and 0.166 (see Sect. 3.7.1). One should always examine the log-likelihood or deviance to be sure that the addition of a new parameter or variable improves the fit.

Now we consider the issue of vegetation and substrate type on the occupancy parameter $\psi$: variables MDH and CDH. Model 3 allows these variables to operate independently, whereas model 8 enforces the equality constraint, MDH = CDH and requires one less parameter to be estimated. All other structural aspects of these two models are the same. Evidence in issues such as this can be provided using an evidence ratio, $E_{3,8} = 0.28938/0.00126 = 230$. Thus, the evidence is strong (my value judgment) that the constraint represents a poor hypothesis. Such evidence ratios do not depend on other models in or out of the set and are useful in contrasting two models that have differing parameterizations.

Because of the interest in mesquite, the relationship between occupancy and the distance to the nearest mesquite was quantified. This was done by computing the evidence ratio between models 6 and 13 and this showed substantial evidence in favor of a relationship ($E_{6,13} = 118$ divided by essentially 0). The estimated slope of the relationship from model 6 was $-0.00075$ with se = 0.00014 and 95% confidence interval of ($-0.0010$, $-0.00048$), further confirming the importance of mesquite to this species of ground squirrel. [Do not be fooled by the low numerical value of the estimated slope ($-0.00075$). It is very small because its associated variable was large (a distance). The importance of this variable is revealed by the relatively small standard error of 0.00014 and the fact that the coefficient of variation is 19%.]

A final aspect of Ball et al.'s (2005) work was to examine the evidence for the SAC habitat model that had been proposed for the management of this ground squirrel. Model $\{\psi, p\}$ most closely represented the proposed habitat SAC model; however, it was ranked third to the last with a model probability of $e^{92.67/2} = 1.33 \times 10^{-20}$. One must conclude that the SAC model was very poor as a basis for management decisions.

## 5.3    Relative Importance of Predictor Variables

In some cases, research is in an early descriptive stage and model selection and valid inference may be constrained by lack of knowledge, small sample size, high dimensionality of the predictor variables, high degree of multicolinearity, and high variability. In such cases, it may be judicious to gain insight into the more important variables from analyzing data from a pilot study and then attempt to collect high quality data on these (few) more important variables. Proper *a priori* hypothesizing and modeling might then focus better on a few variables thought to be important, rather than tackling data on all the variables. This seems like a useful way to approach exploratory data analysis.

### 5.3.1 Rationale for Ranking the Relative Importance of Predictor Variables

Consider a small team of researchers interested in both understanding relationships and making predictions about a response variable $Y$, based on measurements of 15 predictor variables $(x_1, x_2, ..., x_{15})$. There are, in this case, $2^{15} - 1 = 32,767$ possible models, excluding interactions or transformations such as quadratic terms. Rather than gearing up for a huge computer run, the team decides to generate a reasonable subset of variables that seem most important. Thus, an ability to rank the relative importance of the predictor variables might be useful. Then, further research could chase understanding and prediction based on a few variables that rank high. This is not the only approach to making scientific progress under the severe constraints noted, but it is an interesting alternative.

Such ranking can be done with ease if one has some experience with spreadsheets and has a statistical software package that can unthinkingly run "all possible models." In general, I do not recommend running all the models; this is a special case where every variable must be put on an equal footing with the rest for the ranking to be interpretable. Running all the models is an easy way to achieve the balance (fairness) in ranking the relative importance of the predictors. As with AICc model selection, there is no guarantee that any of the predictors are good in some absolute sense; we are merely going to rank them.

### 5.3.2 An Example Using the Cement Hardening Data

The cement hardening data will serve as a handy example with four predictor variables, thus $2^4 - 1 = 15$ possible models. Step 1 is to list all 15 models and their associated model probability (Table 5.7).

Ranking, step 2, is done by merely selecting all the models where $x_i$ appears and summing up the associated model probabilities. Thus, let $i = 1$, then predictor variable $x_1$ appears in the following eight models: {1}, {12}, {13}, {14}, {123}, {124}, {134}, and {1234}. The sum of these eight model probabilities

TABLE 5.7. Summary of the model probabilities for the cement hardening data.

| Model | Probability[a] | Model | Probability[a] |
|---|---|---|---|
| {1} | 0.0000 | {24} | 0.0000 |
| {2} | 0.0000 | {34} | 0.0004 |
| {3} | 0.0000 | {123} | 0.1161 |
| {4} | 0.0000 | {124} | 0.1182 |
| {12} | 0.5670 | {134} | 0.0811 |
| {13} | 0.0000 | {234} | 0.0072 |
| {14} | 0.1072 | {1234} | 0.0029 |
| {23} | 0.0000 | | |

[a]Shown to four decimal places.

is 0.9925. The process is repeated for the eight models where $x_2$ appears, namely models $\{2\}$, $\{12\}$, $\{23\}$, $\{24\}$, $\{123\}$, $\{124\}$, $\{234\}$, and $\{1234\}$.

The results for the four predictor variables are summarized as

| Variable | Sum | Rank |
|---|---|---|
| 1 | 0.9925 | 1 |
| 2 | 0.8114 | 2 |
| 3 | 0.2077 | 4 |
| 4 | 0.3170 | 3 |

In this small example, one might want to focus further work on the two predictors that are ranked high. Note that the use of all possible models gave each variable an equal footing; each variable was in exactly eight models and the sums were all based on eight entries. The method has utility even when $R$ is fairly large (e.g., 20) as standard software can compute the models in a few hours. Then one needs to capture the relevant statistics, compute AICc, the $\Delta_i$ values, and model probabilities, and use a spreadsheet to compute the simple summations.

This ranking procedure will never see heavy use but it is a method worth knowing about when faced with exploratory phases of investigation where dimensionality is high. This ranking approach is most appealing for hypotheses that can be well represented by linear or logistic regression models. The ranking tries to break correlations between and among the predictor variables by having a variable appear on its own and then together with all the other variables. This is an opportunity to determine, via the model probabilities, which variables are related to the response variable and which variables appear to be related, but only because they are correlated with another predictor.

*Ad hoc* procedures such as stepwise regression give a false impression of "importance." Once the algorithm has stopped adding and deleting predictor variables, one might have the final fitted model (where there are 13 predictor variables),

$$\hat{Y} = \hat{\beta}_0 + \hat{\beta}_1(x_1) + \hat{\beta}_6(x_6) + \hat{\beta}_7(x_7) + \hat{\beta}_{12}(x_{12}).$$

Then, one is led to the (incorrect) conclusion that variables $x_1$, $x_6$, $x_7$, and $x_{12}$ are "important" in terms of the response variable. One is compelled to believe that these variables *must* be important because, after all, they are in the final model. Conversely, the remaining variables, $x_2$, $x_3$, $x_4$, $x_5$, $x_8$, $x_9$, $x_{10}$, $x_{11}$, and $x_{13}$ are *surely* not important; after all, if they had been important, they would have been in the final model. Because of the intercorrelations among the predictor variables, such seemingly obvious dichotomies are surprisingly false. Stepwise algorithms do not even identify the second-best model; in fact, no ranking of models is possible using these techniques. Insights such as the one above tend to motivate the use of models beyond just the one estimated to be the best one. There is often substantial *information* in models ranked 2, 3,..., $R$ and it is easy to use this information to allow better inferences to be made from the evidence available. Model based inference ought to be about making inference from all the models in most scientific work. Ranking of the importance of predictor variables is one facet of multimodel inference.

## 5.4   Confidence Sets on Models

Bayesians define an interval in a manner in which they can assert that a single interval contains the parameter with a certain probability (e.g., 0.95). These are often called credible intervals and are easy to understand, whereas the frequentist confidence interval falls back on notions of repeated sampling and the long run coverage at the nominal level (e.g., 0.95). In the real world, there is often little numerical difference between the two types of intervals, but there are worthwhile philosophical differences.

Now consider a set of five science hypotheses represented by five models and a data set with sample size 145. Given this set and the fixed sample size, one of the five models is *the* best model in a K–L sense; we just do not know which one it is (much like an unknown parameter). We can estimate which model is best and the probability that each model $i$ is that actual best model (the model probabilities $w_i$). This thinking leads the way to consider a confidence set on models.

Caley and Hone's (2002) data on bovine tuberculosis in ferrets can be used to illustrate the concept. From Sect. 4.6, we had

| Hypothesis | Model | Model probability |
|---|---|---|
| $H_1$ | $g_1$ | <0.0001 |
| $H_2$ | $g_2$ | <0.0001 |
| $H_3$ | $g_3$ | <0.0001 |
| $H_4$ | $g_4$ | 0.7595 |
| $H_5$ | $g_5$ | 0.2405 |

Summation of the last two probabilities gives 1.0 in this example. We can say probabilistically that the best K–L model is either $g_4$ or $g_5$ (*given* the model set) with virtual certainty in this case. The model set $(g_4, g_5)$ constitutes an approximate 100% confidence set on models. This concept is sometimes useful as an aid in comprehending the meaning of the evidence.

A second example comes from Linhart and Zucchini's (1986) book and deals with prediction and smoothing of weekly data on storm frequency at a botanical garden in Durban, South Africa. They used a combination of logistic regression and Fourier series terms to model weekly storm frequency across 47 consecutive years. The Fourier series terms enter in pairs (sines and cosines); thus, models are nested and $K$ jumps by 2 from model to model. I will avoid other complications (e.g., overdispersion) and provide a summary of the quantities available:

| Model $i$ | $\Delta_i$ | Model probability |
|---|---|---|
| 1 | 39.04 | <0.0001 |
| 2 | 1.64 | 0.2141 |
| **3** | **0.00** | **0.4861** |
| 4 | 1.49 | 0.2305 |

(continued)

| | | |
|---|---|---|
| 5 | 3.98 | 0.0665 |
| 6 | 10.59 | 0.0024 |
| 7 | 17.24 | <0.0001 |
| 8 | 24.51 | <0.0001 |
| 9 | 33.29 | <0.0001 |

Model 3 with six parameters is the best with probability 0.4861 flanked by model 2 and model 4, with probabilities 0.2141 and 0.2305, respectively. The sum of these three probabilities gives 0.93, giving an approximate 95% confidence set. Including the probability for model 5 gives a 99.7% confidence set. These sets are only approximate but allow the notion of confidence intervals for parameters to be extended to confidence sets for models. Here, models other than 2, 3, and 4 lie outside a 93% confidence set. Clearly, models 1, 7, 8, and 9 lie well outside either set.

The use of confidence sets on models is occasionally useful, particularly when the models are nested. These sets can help understand subsets of models that have reasonable plausibility.

## 5.5   Summary

The idea of making formal inductive inferences from an array of *a priori* models is compelling. Given a choice of using one model where there is uncertainty concerning its rank and using all the models in the set, I think people would prefer the latter. Multimodel inference seems generally desirable. The curious thing is that multimodel inference is computationally easy. In the future, it seems likely that additional approaches will be developed to allow inference from multiple models.

## 5.6   Remarks

A good discussion of model averaging is given by Hoeting et al. (1999). Their paper is written in a Bayesian setting, but the review of the general approach is good reading.

Anthony et al. (2006) provide the results of an enormous research program on the Northern Spotted Owl (*Strix occidentalis caurina*). The analysis of these data was done under a multimodel inference paradigm due partly to the litigious nature of the long-term controversy over cutting old growth forests and its effects on spotted owls and other conservation concerns. Model averaging can have substantial values when the science issue involves controversy (see Hoeting et al. 1999; Anderson 2001).

Chatfield's (1995b) extensive paper is excellent on several important, but perhaps subtle, issues; in particular, problems that arise when using the data to both select a model and then make inferences from that selected model.

These concepts are largely handled by the information-theoretic approaches for many classes of problems.

Breiman (1992) offered the term "quiet scandal" when estimates of precision are presented without a variance component for model selection uncertainty.

Burnham and Anderson (2002:Chap. 5) provide the results from a number of MC simulation studies showing the poor confidence interval coverage of estimators when model selection uncertainty is ignored. They simulated binomial data (10,000 replicates) from a simple age-specific survival model with ten age classes with sample size of 150 subjects. The model set allowed estimates of survival probability up to some age, whereas the remaining age classes were pooled, as is often done when the number of survivors dwindles. Inference was made from the best model and confidence interval coverage was poor when only the sampling variance was used as a measure of precision: mean coverage was 81.3%, ranging from a low of 63.0% to 95.9%. In contrast, when a variance component for model selection uncertainty was added, coverage averaged 95.0%, ranging from 90.6% to 97.7%.

Various approaches to model selection began to appear in the technical literature since computers became available in the 1960s. Prior to that, one was happy to obtain the MLEs and covariance matrix for a single model as calculations were laborious and had to be done by hand. Procedures such as stepwise regression filled an important void and saw heavy use. Only in the past 15 years have people begun to ask about the statistical properties of the selected model (be it from stepwise, Mallows' $Cp$, AICc, or whatever). It became clear that the estimators used in the selected model had confidence interval coverage below the nominal level. This limitation was caused because model selection uncertainty was not embedded into estimates of precision (Chatfield (1995b) covers this issue and provides insights into problems with data dredging).

Statistical software packages could be much more useful if they treated sets of models, given a data set, rather than treating individual models in isolation. I am aware of only two major software packages that take this approach: program MARK (White and Burnham 1999) and Distance 5.2 (Thomas et al. 2006). Both of these packages are freeware; however, neither is general purpose statistical package.

Predictor variables in linear and nonlinear regression are often correlated and this has its consequences. The cement hardening data will serve as an example where the variables $x_2$ and $x_4$ had a correlation coefficient of −0.973. One advantage of using AICc is that both of the variables are retained in the analysis. Thus, models {12}, {14}, and {124} were three of the best four models, with model probabilities of 0.567, 0.107, and 0.118, respectively. Although $x_4$ is not the best of the pair, perhaps this variable is very much less expensive to measure; thus, it should not be lost from the results. There are several ways to handle correlated variables, including a simple geometric mean of the members of the pair, thus reducing two variables to one. If several similar variables have high correlations, one can perform a principal components analysis (PCA) and hope that most of the variation is contained in the first 1–2 components; however, issues of interpretability often arise.

Ranking the relative importance of variables is a more sound way to try to identify important variables from a large set. In the past, people have used some sort of statistical test to sequentially weed out "nonsignificant" variables and this approach has poor properties (the multiple testing problem to mention only one issue).

I hope the reader is gaining an appreciation for how bad *ad hoc* procedures such as stepwise regression can be, even in routine situations where the assumptions are fairly well met. Although stepwise methods are still being taught routinely, they are a poor basis for model based inference in a linear models setting (see McQuarrie and Tsai 1998).

Guidelines have been published outlining the quantities that should often appear in publications (Anderson et al. 2001b). In the ground squirrel example, the issue surrounding model 2 could not have been uncovered had the value of the log-likelihood (or the deviance) not been published.

## 5.7    Exercises

1. The first exercise in Chap. 4 dealt with the data in bill lengths in Darwin's finches. Would you employ model averaging the estimates of $\beta_1$ in this case? Why? Why not? Would you do any model averaging in this example? Should model selection uncertainty be incorporated into estimates of precision in this example?

2. Review Table 5.5 from the study of Palm Springs ground squirrels. Your colleague provides you with the evidence ratio $E_{5,8} = 2.65$. Write a concise paragraph explaining the biology implied by this result.

3. When faced with many predictor variables in linear or logistic regression one must often try to reduce the dimensionality by various means. One approach has been to perform a principal components analysis (PCA) on the $X$ matrix. Then, the regression is on PCI, PC2,..., rather than on the original variables $x_1$, $x_2$.... What are the advantages and disadvantages to this approach? (advanced question)

4. A nonparametric bootstrap might be used in model selection. Outline this approach in a one-page report and offer a critique. (advanced question)

5. We learned in Chap. 2 that information was additive. How might this fact be exploited using the $\Delta_i$ values? (advanced question)

6. Review the paper by van Buskirk and Arioli (2002) and consider ways in which their model set might evolve to the next level, given their results.

7. Bortz and Nelson (2006) studied HIV infection dynamics that surely gives some insight into modeling complex system using state-of-the-art quantitative methods. Readers with a background in various types of differential equations and mixed effects modeling should read this paper.

   a. What is gained by thinking that the "penalty term" in AIC, AICc, and TIC is a measure of "complexity"?

   b. They seem to favor an information criterion termed ICOMP. Can you determine the rationale for this choice?

   c. Is it not surprising that $K$ is so small for the models they evaluate?. Why might this be?

# 6
# Advanced Topics

Kenneth P. Burnham (1942–) Dr. Burnham received a B.S. degree in biology from Portland State University and his M.S. and Ph.D. degrees in mathematical statistics in 1969 and 1972 from Oregon State University. He has worked at the interface between the life sciences and statistics in Maryland, North Carolina, and Colorado. He has made long strings of fundamental contributions to the quantitative theory of capture–recapture and distance sampling theory and analysis. His contributions to the model selection arena and its practical application have been profound. He was selected as a Fellow by the American Statistical Association in 1990 and promoted to the position of Senior Scientist by the U.S. Geological Survey in 2004. He has a long list of awards and honors for his work, including the Distinguished Achievement Award from the American Statistical Association and the Distinguished Statistical Ecology Award from INTERCOL (International Congress of Ecology). He has just become an elected member in the International Statistical Institute. Ken (left) is shown with Hirotugu Akaike at the 2007 Kyoto Laureate Symposium. Photo courtesy of Paul Doherty and Kate Huyvaert.

# 6.1    Overdispersed Count Data

Statistical methods are often based on the "iid" assumption: independent and identically distributed data. This assumption is nearly always made in application (time series and spatial models are exceptions); however, in reality, data are often somewhat dependent and not identically distributed. These conditions fall under the concept of *overdispersion*. Count data (zero and the positive integers stemming from some count) are often said to be "overdispersed." There are two issues here. First, overdispersion is a property of the data, not a model; however, overdispersion can be modeled. Second, overdispersion can be modeled as either a lack of independence or parameter heterogeneity. There are a variety of specialized approaches to attempt to deal with overdispersed data, most are at an advanced level and specific to certain problem types. A simple method often serves to lessen the problem with overdispersed count data and it will be introduced in this section.

## 6.1.1    Lack of Independence

Flipping new pennies and observing the binomial outcomes (i.e., heads or tails) nicely illustrates independence of the outcome from flip to flip. However, count data in the life sciences often have some degree of dependence. Husbands and wives may not be independent with respect to some condition. Individuals in small groups of tadpoles along a mud bank probably die or survive with some degree of dependence within a group. If one tadpole in a group dies it may be that many others die at about the same time and for the same underlining cause. The analysis of count data of litter mates, breeding pairs, schools of fish, and pods of whales should always be suspected of having some degree of dependence. If such count data are analyzed as if they were independent, then the sampling variances tend to be too small (underestimated), giving a false sense of precision (e.g., confidence intervals are too narrow).

## 6.1.2    Parameter Heterogeneity

Overdispersion can also arise as most statistical methods rely on the concept of parameter homogeneity. Although 200 new pennies may each have essentially the same probability of a head, it is clear that 200 laboratory mice are somewhat variable, almost regardless of the trait of interest. This individual variation leads to what is termed "parameter heterogeneity" and this violates the *iid* assumption. Again, the effect of such heterogeneity, if the data are analyzed under methods that assume parameter homogeneity, is again the underestimation of sampling variances. There may be substantial bias in parameter estimates in some isolated cases.

### 6.1.3   Estimation of a Variance Inflation Factor

Overdispersion causes the estimated theoretical variances and covariances to be biased low; thus, a first-order approach is to "inflate" these up to a nominal level. This is a simple and often effective procedure. First, we focus on a robust global model; a model with plenty of structure. Here we must assume that there is no structural lack of fit and, therefore, lack of fit can be blamed on overdispersion. This is a strong assumption and one risks the situation where some of the lack of fit is a structural inadequacy of the model and not overdispersion. In this case, the covariances would be inflated (and this might be beneficial) when, in fact, some bias is likely due to inadequate structural modeling.

If there is little reason to suspect some dependence among observations based on counts, then perhaps one should ignore the issue. However, if there is biological reason to suspect overdispersion, then an overdispersion parameter $c$ can be estimated,

$$\hat{c} = \chi^2 / \mathrm{df},$$

where $\chi^2$ is the usual goodness-of-fit test statistic based on the global model and df is the degrees of freedom for the test. The overdispersion ($c$) parameter is also called a *variance inflation factor*. Under the *iid* assumption, $c \equiv 1$. In biological data on counts one often sees $\hat{c}$ in the 1–3 range. Fish in schools, insects in colonies or swarms, or snakes in dens can have overdispersion parameters substantially higher than 4–5. As $\hat{c}$ gets large one must worry that there are structural issues with the model and these are being incorrectly cast as overdispersion.

### 6.1.4   Coping with Overdispersion in Count Data

---

**Coping with Some Dependence**

Used carefully, the estimation of an overdispersion parameter can adjust the analysis in the face of some degree of dependence and parameter heterogeneity. If overdispersion is thought to be an issue and an estimate of the overdispersion parameter is available, e.g., $\hat{c}$, then three things should be done in the analysis (the order is not important):

1. The log-likelihood of the parameters $\theta$, given the data and the model, should be computed as

$$\frac{\log\left(L(\theta \mid x, g_i)\right)}{c},$$

therefore, model selection should use the following modified criterion

$$\mathrm{QAICc} = -\left[2\log\left(L(\hat{\theta})\right)/\hat{c}\right] + 2K + \frac{2K(K+1)}{n-K-1}$$

---

2. The number of parameters ($K$) is now the number of parameters (the dimension of $\theta$) in the model, plus 1 to account for the estimation of the overdispersion parameter, $c$

3. The variance–covariance matrix should be multiplied by the estimated overdispersion parameter, $\hat{c}$ (i.e., $\hat{c}$ (cov($\hat{\theta}_i$, $\hat{\theta}_j$) for all the models. Thus, $c$ is used to actually inflate the variance and covariances. Alternatively, standard errors are inflated by the square root of $\hat{c}$.

Once an estimate of the overdispersion parameter has been made from a global model, it is used for all the models in the set (i.e., the three steps outlined above). If $\hat{c} < 1$, then it is rounded up to 1 and no adjustment is made in any of the above quantities. The notation QAICc stems from the concept of quasi-likelihood from a well-known paper by Wedderburn (1974).

The log-likelihood is adjusted in an intuitive way. Usually, the log-likelihood contains all the information in the sample data, given the model and assuming independence. When, instead, there is some dependence, a log-likelihood that assumes independence exaggerates the amount of information in the data. Thus, division by the estimated overdispersion coefficient correctly adjusts the log-likelihood for the degree of dependence reflected in the data.

Highly dependent data have considerably less information and $\hat{c}$ is needed to adjust for the dependence. Assuming everything else is constant, highly dependent data reflect less precision for parameter estimates and selected models with fewer parameters or less structure.

### 6.1.5  Overdispersion in Data on Elephant Seals

Pistorius et al. (2000) evaluated hypotheses concerning age- and sex-dependent rates of tag loss in southern elephant seals (*Mirounga leonina*) by considering four models. There was belief that these data were overdispersed due primarily to parameter heterogeneity. Burnham and Anderson (2001) made use of these data as an example to explore these issues further. They performed a goodness-of-fit test (*TEST2*, Burnham et al. 1987) on these data, partitioned by gender. The results were

| Quantity | Males | Females | Combined |
|----------|-------|---------|----------|
| $\chi^2$ | 157.20 | 97.92 | 255.12 |
| df | 77 | 84 | 161 |

giving $\hat{c} = 255.12/161 = 1.58$. This suggests some minor to moderate overdispersion and it is likely to be worthwhile to inflate the variances and covariances and alter the deviance.

Thus, QAICc was used, whereby the deviance was computed as $-2\log(\mathcal{L}(\phi))/\hat{c}$, the parameter count ($K$) was increased by 1 for the estimation of the variance inflation factor, and the covariance matrix for the four models was multiplied by $\hat{c} = 1.58$. Pistorius et al. (2000) used the bootstrap to obtain estimates of sampling variance and they found the empirical support to be for the models where tag loss was sex- and age-dependent ($w_{best} = 0.82$) or just age-dependent ($w_{second} = 0.18$).

As dependence in the data increases, QAICc will tend to select less rich models (i.e., fewer parameters and less structure). This result follows because there is less information when some dependence is present in the data. In a sense, the "effective" sample size is less than $n$. Underdispersion seems hard to imagine; I have not seen this in my experience.

A reviewer brought up the question of independence in time series and spatial modeling problems. Here, the "response variable" is correlated in time or space. Thus, it is the model that attempts to handle the dependencies in time or space (see Renshaw 1991). If successful, the "residuals" will be uncorrelated.

## 6.2   Model Selection Bias

Technical difficulties can arise when using data to both select a good model and estimate its parameters. Chief among these is the subtle but very important issue of *model selection bias*.

It is difficult for most of us to understand model selection bias because in our regression classes we learned that, given the model, the $\beta_i$ were unbiased, normal, and have minimum variance. This is all true, *given the model* and its underlying assumptions. In the real world, *the* model is not *given* to us, we must use some analytic approach to select a good model from the data. Again, issues arise when the same data set is used to both select a model and estimate its parameters.

### 6.2.1   Understanding the Issue

This issue can best be understood in terms of linear or logistic regression. I begin by considering the linear regression function, and for simplicity I will assume all the $\beta_i$ are positive (and then the discussion relates to overestimation),

$$E(Y) = \beta_0 + \beta_1(x_1) + \beta_2(x_2) + \beta_3(x_3).$$

We will define $x_1$ as a "dominant variable" because its relationship to the response variable ($Y$) is quite important. Here, we might expect $\beta_1/se(\beta_2) \approx 3$ or 4. Nearly all methods would select this variable to be important (see Miller 2002). The variable $x_2$ is, in fact, somewhat important, but its relationship to the response variable is a bit weak. Perhaps $\beta_2/se(\beta_2) \approx 1$. Finally, variable $x_3$ is tenuous at best. Here, perhaps $\beta_3/se(\beta_3) \approx 1/4$. The epidemiologist Michael Thun noted, "… you can tell a little thing from a big thing. What's very hard to do is to tell a little thing from nothing at all" (Taubes 1995). This is the concept of tapering effect size (Sect. 2.2.5). Here, $\beta_3$ is nonzero and reflects a very weak effect.

Assume we use the linear model above to generate 1,000 data sets where the $\beta_i$ parameters are known, then we can use a model selection approach (e.g., stepwise, AICc or BIC) to find the best model for each of the data sets. Variable $x_1$ will likely be selected in virtually all of the 1,000 data sets, while variable $x_2$ might be selected in perhaps half of the data sets. That is, too little information is contained in many of the data sets and, in view of parsimony, the importance of $x_2$ and its $\beta_2$ is not picked up. More interestingly, it might be that variable $x_3$ and its parameter $\beta_3$ is selected in only a few (i.e., 3–6%) of the data sets.

Now we must ask what are the properties of the estimator $\hat{\beta}_3$ *when it is in the selected model*? Large bias is the answer! The bias arises because about the only time $x_3$ is in the selected model are cases where it is overestimated. If $\hat{\beta}_3$ is near the actual value $(\beta_3)$, then the variable does not appear in the selected model. It is only when it happens to be overestimated that it is selected and in the best model. Thus, when one averages across (the few) models where $\hat{\beta}_3$ appears, it is far too large. Thus, a large bias is present and it is this that is called *model selection bias*. The issue extends to variables that are exactly unimportant; i.e., where $\beta = 0$. Occasionally, $\hat{\beta}$ will be large and the inference will be that this variable is important. Model selection bias is not an easy concept but it is both common and important. Model selection bias can often be in the 10–80% range, but can be far more serious (see Miller 2002, for some examples).

When a given data set gives rise to a substantial overestimate, a standard Wald test, $t = \hat{\beta}_3 / \hat{se}(\hat{\beta}_3)$, would be "highly significant" and $x_3$ and its parameter estimate of $\beta_3$ would be in the model. In this case, the numerator $(\hat{\beta}_3)$ is biased high and the denominator $(\hat{se}(\hat{\beta}_3))$ is biased low, yielding an unreliable test result. AICc and other approaches have this same strong tendency but is harder to demonstrate in an analogous way.

When one has 1–2 dozen predictor variables (i.e., 4,095–16,777,215 models), the opportunity for large biases due to model selection are enormous and the probability of several spurious effects quickly goes to 1. Model selection bias is subtle but its effects are widespread and little understood by many people working in the life sciences.

Model selection bias should be a worry in applied data analysis because the analyst has no way of knowing, from the analysis of a single data set, which parameters might be very much overestimated and which have little bias. In fact, the inference that $x_3$ is very important is largely spurious. This problem is compounded in that the estimated sampling variances are too low (underestimated), giving a false sense of high precision. Driving this issue is the concept of tapering effect sizes that seem omnipresent in the real world.

## 6.2.2   A Solution to the Problem of Model Selection Bias

Model averaging offers a solution to the problems of model selection bias (P. Lukacs and K. Burnham, personal communication). This approach applies

TABLE 6.1.    Model averaging as a means of reducing model selection bias. The model probabilities are shown at the far right.

| | | | | | |
|---|---|---|---|---|---|
| 1 | $\hat{Y} = \hat{\beta}_0$ | $+\hat{\beta}_1 X_1$ | $+\hat{\beta}_2 X_2$ | $+\hat{\beta}_3 X_3$ | 0.15 |
| 2 | $\hat{Y} = \hat{\beta}_0$ | $+\hat{\beta}_1 X_1$ | $+\hat{\beta}_2 X_2$ | | 0.35 |
| 3 | $\hat{Y} = \hat{\beta}_0$ | $+\hat{\beta}_1 X_1$ | | $+\hat{\beta}_3 X_3$ | 0.10 |
| 4 | $\hat{Y} = \hat{\beta}_0$ | | $+\hat{\beta}_2 X_2$ | $+\hat{\beta}_3 X_3$ | 0.05 |
| 5 | $\hat{Y} = \hat{\beta}_0$ | $+\hat{\beta}_1 X_1$ | | | 0.25 |
| 6 | $\hat{Y} = \hat{\beta}_0$ | | $+\hat{\beta}_2 X_2$ | | 0.10 |
| 7 | $\hat{Y} = \hat{\beta}_0$ | | | $+\hat{\beta}_3 X_3$ | 0.00 |

to model parameters, not predictions that were covered in Chap. 5. The approach is a type of shrinkage (see below) estimation using model averaging. We will use the case outlined above where $x_1$ was a dominant variable, $x_2$ was far less important, and $x_3$ was barely nonzero. The models with their associated model probabilities are shown in Table 6.1.

There is a "balancing" such that each of the $\beta$ slope parameters occurs in 4 of the 7 models. Such balancing can be done is one of several ways for many problems, but each parameter should be allowed an equal footing. It is often sufficient to list "all possible models" as a way to achieve the needed balance.

Notice that all the models with the dominant variable tend to have high weights (model probabilities). In contrast, the model with only $x_3$ has virtually no weight. A robust estimate of each of the 3 $\beta$ parameters can be made in the usual manner,

$$\hat{\bar{\beta}}_1 = \sum_{i=1}^{7} w_i \hat{\beta}_{1i} \quad \hat{\bar{\beta}}_2 = \sum_{i=1}^{7} w_i \hat{\beta}_{2i} \quad \text{and} \quad \hat{\bar{\beta}}_3 = \sum_{i=1}^{7} w_i \hat{\beta}_{3i},$$

but when a regression parameter does not appear in a model, it is assigned a value of 0. The fact that a parameter does not appear in a model *implies* it has a zero value, and so this ought not seem too surprising upon consideration. Note, in all cases, the model probabilities sum to 1

$$(\text{i.e.,} \sum_{i=1}^{7} w_i = 1)$$

For example, the model averaged estimator for $\beta_3$ is $\hat{\bar{\beta}}_3$ computed as a simple weighted average, where zeros (in bold) are assigned for models where the parameter does not appear (see Table 6.2).

That is, $\hat{\bar{\beta}}_3 = \sum_{i=1}^{7} w_i \hat{\beta}_{3i} = 0.610$. This estimate is below the MLEs as it has been "shrunk."

The fitted equation is a single equation where the parameters have all been model averaged,

$$\hat{Y} = \hat{\bar{\beta}}_0 + \hat{\bar{\beta}}_1 x_1 + \hat{\bar{\beta}}_2 x_2 + \hat{\bar{\beta}}_3 x_3.$$

TABLE 6.2.    Computing the model averaged estimate of $\beta_3$.

| Model | Model weight | MLE of $\beta_3$ | Product |
|---|---|---|---|
| 1 | 0.15 | 1.73 | 0.260 |
| 2 | 0.20 | **0.00** | 0.000 |
| 3 | 0.25 | 0.83 | 0.208 |
| 4 | 0.10 | 1.42 | 0.142 |
| 5 | 0.20 | **0.00** | 0.000 |
| 6 | 0.10 | **0.00** | 0.000 |
| 7 | 0.00 | 1.08 | 0.000 |
| Sum | 1.00 | | 0.610 |

This single equation allows robust predictions to be made, based on each of the regression parameters having been model averaged. This procedure "shrinks" estimates toward zero, and greatly lessens the bias due to model selection. Finally, note that predictions made using this model (above) are identical to those made by making a prediction from each model and then model averaging these. The two approaches are equivalent for linear models.

MC simulations have been carried out to suggest this simple approach is very effective (Lukacs et al. unpubl ms.). Here it is important to use the unconditional variance to account for model selection uncertainty.

Use of "all possible models" is a poor strategy in general; in this specific case using all the models is a simple way to impose a balance and put each variable on an equal footing (e.g., each variable appeared in exactly 4 of the 7 models in this example). There are other ways to maintain this balance; for example, in the data on hardening of Portland cement the single variable models were ruled implausible. Thus, one could get the shrinkage estimates from 11 of the models, rather than using the full set of 15 models and still achieve the needed balance.

Freedman's (1982) paradox is largely resolved using this model averaging approach. The combination of shrinkage model averaging and unconditional variances help guard against spurious results. Freedman (1983) concluded, "To sum up, in a world with a large number of unrelated variables and no clear *a priori* specifications, uncritical use of standard methods will lead to models that appear to have a lot of explanatory power." It seems that this type of model averaging will appear to be useful in lessening these issues. Additional research on this matter will be useful in application. Lukacs et al. (unpublished manuscript) have completed some simulations using logistic regression and found performance to be good in this case. While I do not recommend wholesale data dredging via "all possible models," I believe this type of model averaging will help avoid serious model selection bias when faced with many predictor variables, little science theory, and small sample sizes.

## 6.3   Multivariate AICc

When one is performing multivariate analyses (e.g., multivariate regression or factor analysis), a slightly altered model selection criteria must be used. Terminology can be confusing; here I am addressing the case where there are more than one response variables (multivariate regression vs. multiple regression – both will typically have several predictor variables). The altered criterion is from Fujikoshi and Satoh (1997) (also see Bedrick and Tsai 1994),

$$\text{AICc} = -2\log(\mathcal{L}) + 2K + 2\frac{K(K+v)}{(np-K-v)},$$

where $n$ is sample size on observations on each of the $p$ variables and $v$ is the number of distinct parameters estimated in the covariance matrix ($K$ includes $v$). Also see Sclove (1994b), McQuarrie and Tsai (1998:147–149), and Burnham and Anderson (2002:424–426), for additional discussion. In the univariate case, $p = 1$ and $v = 1$ and the criterion above reduces to AICc for univariate cases.

Model selection criteria for multivariate analysis represent an active research area. Recent work includes Siotani and Wakaki (2006) and Seghouane (2005, 2006). It is certainly prudent to work closely with a statistician with expertise in multivariate analysis before going too far with an analysis of complex multivariate data.

## 6.4   Model Redundancy

Occasionally the model set contains two or more models that are inadvertently the "same." This condition is termed *model redundancy*. Model redundancy does not cause problems with $\Delta_i$ values, model likelihoods, or evidence ratios, but the model probabilities ($w_i$) are affected.

Model redundancy can arise as a mistake or the result of carelessness. Alternatively, a team member might suggest a model where a transition probability ($\psi$) is modeled as a function of distance ($d$) from a source as

$$\psi(d) = \frac{\exp\{\beta_0 + \beta_1(d)\}}{1 + \exp\{\beta_0 + \beta_1(d)\}},$$

while another person suggests the model

$$\psi(d) = \frac{1}{1 + \exp\left[-\{\beta_0 + \beta_1(d)\}\right]},$$

These models are actually the same and will have the same $\log(\mathcal{L})$ values. Many models have a chameleon-like form and it is sometimes easy to have 2 or more models with different forms that are actually the same model.

Model redundancy can arise when using semiparametric models (Buckland et al. 1997) where the number of parameters is not fixed, but rather enter from the results of model selection. Consider the parameters in a series of Fourier series being used to smooth and make predictions of weekly storm events discussed in Sect. 5.4. These were nested models where 2 parameters were gained from model to model. Thus, $g_1$ had 2 parameters, $g_2$ had 4 parameters, $g_3$ had 6 parameters, and so on. Model selection indicated that model $g_2$ with 4 parameters was satisfactory. It is then possible that, without realizing it, the analyst believes he has 4 models with 2, 4, 6, and 8 parameters, respectively. However, the last 3 of the 4 models actually have but 4 parameters (because the additional parameters in models 3 and 4 were not needed (dropped).

Model redundancy can arise in subtle ways and steps should be taken or risk the possibility that the model probabilities will be affected, leading perhaps to inferences that are needlessly poor. The usual effect is that one model gets too much weight. The first and most effective solution is to identify the redundant models and delete them from consideration. This will mean the model probabilities ($w_i$) must be renormalized, but this is trivial to do.

The alternative solution is to cast the redundant models into a subset and assign the models in this subset an appropriate "weight." For example, let there be 6 models in the set and 2 are found to be redundant (let these be models 5 and 6 for illustration). The solution is to compute the model probabilities using the expression

$$w_i = \frac{\exp\left(-\tfrac{1}{2}\Delta_i\right)\xi_i}{\sum\limits_{r=1}^{R}\exp\left(-\tfrac{1}{2}\Delta_r\right)\xi_r},$$

where the $\xi_i$ are 1/5, 1/5; 1/5, 1/5; and 1/10, 1/10. This simple device merely gives the two redundant models 1/2 their usual weight. Note, the $\Sigma\xi_i = 1$ as it must.

## 6.5    Model Selection in Random Effects Models

This book has been about models that can be termed "fixed effects" models. There is an interesting class of models falling under a general classification of "random effects models;" roughly alternative names include "variance component models," "random coefficient models," and "hierarchical models" (see Vonesh and Chinchilli 1997; Shi and Tsai 2002; McCulloch 2003; and Gurka 2006, for additional details). Because of the design of the data collection, such procedures often allow an inference that is wider in scope than with fixed effects models. These approaches allow separate estimation of a component of variance due to sampling (var($\hat{\theta} \mid$ model, $\theta$)), distinct from a process variance component ($\sigma^2 = $ var($\hat{\theta}$)). Here process variation might be temporal or spatial. In a sense, the goal of random effects models is the estimation of population means ($\mu$) and process variances ($\sigma^2$) and this can often occur without models with a large number of parameters.

Other useful approaches allow shrinkage estimators; these are estimators that attempt to shrink estimates toward their mean value and the amount of shrinkage depends on the relative magnitude of the sampling variance to the process variance. The number of estimable parameters in shrinkage approaches may not be an integer. Such estimators, as a set, can have smaller mean squared errors (MSEs) than MLEs for the same data. Then there are "mixed" models that allow for both fixed and random effects and these too are being used often. All of these approaches exist at a somewhat advanced level and are seeing increased use in many applied fields in the life sciences. There is a large literature on this important class of models (see Gurka 2006, and references therein).

Model selection for random effects and mixed models can be done under a Kullback–Leibler information framework without modification (assuming the proper likelihood is used and the "number" of parameters is available and correct). Here, standard software provides things such as the RSS or the maximized log-likelihood and AICc can be easily computed by hand. Several software packages now output AIC; however, rarely is AICc provided. The key here is to be sure the correct likelihood is in place.

More complicated hierarchical models (e.g., multilevel hierarchical effects) are very well suited for Bayesian Markov Chain Monte Carlo methods (Gelman et al. 2003; Givens and Hoeting 2005). This is a class of models where Bayesian approaches have a distinct edge over other methods; however, methods based on "h-likelihood" may eventually provide another alternative (Lee et al. 2006). There is a large Bayesian literature on these methods and many applications are beginning to appear in journals in the applied sciences. Model selection in these classes of models seems to rely on DIC, the deviance information criterion (see Spiegelhalter et al. 2002). DIC is implemented in programs BUGS and WINBUGS and is seeing heavy use. DIC has Bayesian roots, comes easily from the MCMC algorithm, and is largely AIC-like in its goals and properties.

## 6.6   Use in Conflict Resolution

The focus of this book as been on science philosophy and the provision of empirical evidence for *a priori* science hypotheses. This final section mentions the use of information-theoretic methods in the resolution of certain types of conflicts. The key to this application is a "protocol" that is jointly developed as an *a priori* template to guide the resolution of the conflict. This approach assumes that data and data analysis are central to the resolution of the conflict and, therefore, defines a relatively narrow region of potential application.

In all subdisciplines in the life sciences, there are conflicts and controversies and many of these concern technical issues (e.g., does smoking cause lung cancer? does exposure to low levels of lead lower IQ?). Many such controversies exist when alternative economic, social, or legal outcomes hinge on scientific results.

Often as a controversy starts to brew one individual or party will take a partition of the available data, analyze it in a way that suits him, develop the results, and show them to the larger group, expecting them to yield to his position. Unimpressed, others in the group question the choice of the data used, the methods, and the "obviously biased" result. These people then retreat to use the data they feel can be justified, choose their own analysis method, develop what they believe to be *the* results and again expect the larger group to yield to their conclusion. By then personalities are on edge and the controversy may begin to enlarge and become personal. At this stage, the controversy may continue to brew over long time periods or be headed for the courts. It seems better to try to get agreement on the substantive, technical issues; then if the courts get involved it is over the less tangible political issues, but hopefully based on good science.

Here it seems important to clearly separate the science issues (does smoking cause lung cancer or not?) from the management or political implications and the related value judgments (e.g., smoking causes cancer; therefore, ban all smoking products or, at least, tax smoking products heavily). The material to follow suggests a protocol for resolving the science issues in a controversy. Science should ideally provide a uniform result; science results should not align with sponsorship or employment of the scientists. The material in this section is taken largely from Anderson et al. (1999).

---

**Evidence in Conflict Resolution**

The overall theme in using information-theoretic methods in the resolution of scientific controversies is to replace the *a priori* set of science hypotheses ($H_i$) with an *a priori* set of "stakeholder" positions ($S_i$). In both cases, similar issues are important: careful definition of the problem, good data, sound analysis methods, quantified evidence, and synthesis, usually followed by value judgments.

The goal is then to examine the empirical support for each position with an *a priori* understanding as to what is expected under different outcomes.

---

## 6.6.1 Analogy with the Flip of a Coin

The protocol is patterned after the flip of a coin to decide a course of action. In a coin flip, there are numerous issues that must be decided and agreed upon prior to the flip, such as (1) who flips the coin? (2) should the coin land on the floor or on the back of one's hand? (3) who gets to choose "heads" or "tails"?

(5) who gets to flip? and (6) most importantly, what is the exact action to be taken if the coin comes up "heads"?

Take the example of Mary and John deciding who will pick up the tab for lunch. Mary and John debate the preliminary issues and mutually agree that John will flip a single coin, that it must land on the floor, and that Mary chooses "heads" while the coin is in the air, and that this outcome means that John must buy lunch (i.e., "heads" = John buys and "tails" = Mary buys). There is clear, deliberate agreement on these *a priori* issues. These issues represent the agreed upon protocol.

The key to the coin flip protocol is that it is clearly unfair or unethical for one party to change their choice *after* the flip! That is, Mary cannot expect to switch to a position "if it is 'tails', John must cover the expenses" *after* the coin has landed and the outcome noted. Similarly, she cannot decide not to play, once the coin has landed and she has lost. The parties can argue about the preliminaries, but once these are agreed upon and the coin is "flipped," they cannot argue the outcome (e.g., "heads," therefore John must buy). Of course, if agreement cannot be found on, say, who flips the coin, then the matter of lunch expenses must be settled in one of many other ways (i.e., the player decides not to play). A player can withdraw with honor anytime during the development of the protocol. Football and many other sporting events use a coin flip with specific protocols as part of achieving "fair play."

## 6.6.2   Conflict Resolution Protocol

The conflict resolution protocol rests on the *a priori* agreement by all parties on

- The questions to be addressed
- The data to be analyzed,
- The specific data analysis methodologies
- Who performs the analysis
- What outcomes provide evidence for which stakeholder position (to favor one side or the other or remain ambivalent)
- How these outcomes will be announced, reported, or reviewed

The fundamental idea is to argue points and eventually agree on the relevant data (which cannot be changed after results are known), an analysis protocol, and then agree on the interpretation of the results within certain limits. This final point (interpretation) attempts to avoid any ambiguity where both sides argue, after the analysis has been completed, "that proves what I said."

Management implications (the nonscience) based on empirical results (the science) may often be open to discussion and intense debate. The protocol advocated deals only with the science of the matter. The protocol outlined provides a useful, general framework to deal with the synthesis and analysis of empirical data where decisions are to be based on empirical data ("the best available science").

Often, synthesis of empirical data for management decisions comes from disparate sources with differing analytical methodologies and interpretations.

Such reviews often list tables of results from different published and unpublished studies from which conclusions are made. However, this approach is often hampered by the different analytical methods used in the separate studies. The approach suggested here differs from those approaches in that there is a deliberate attempt to unify separate studies under a single analytical philosophy and framework that was agreeable, *before the results were in*, to all parties involved. This procedure allows the synthesis of empirical data to have greater scientific credibility and clearly demands consensus among the parties involved regarding methods of data collection and analysis.

Using this protocol might often avoid acrimonious and expensive judicial hearings to arbitrate controversies. In such hearings, both parties present evidence to support only their own, often vested, position. These hearings often aggravate controversy, widen disagreements, and confuse the evidence. While the judicial model has several advantages, I suggest that scientists and managers should attempt an objective resolution of scientific issues, rather than turn over these technical tasks to opposing teams of attorneys and a judge.

Often, relevant data, proper analysis methods, and the interpretation of results in terms of management are disputed by the parties. Issue resolution requires numerous features in our protocol, including involving outside parties with minimal vested interest in the outcome, and *a priori* consensus on directions to proceed. Such issues would benefit from the application of the protocol, once those directions are established. I stress that there is a good deal of flexibility in the application to other situations as long as the philosophical core remains intact.

### 6.6.3  A Hypothetical Example: Hen Clam Experiments

Anderson et al. (2001) illustrated the use of information-theoretic approaches using a hypothetical experiment to examine the effects of a chemical on monthly survival probabilities of the hen clam (*Spisula solidissima*). A registered chemical (*Llikmalc*) was applied aerially across aquatic habitats for mosquito control and controversy arose over its unintended effects on other aquatic organisms. The hen clam became the subject of conflict between (a) the manufacturer and distributor of *Llikmalc*, (2) the state regulatory agency, and (c) an environmental group.

The protocol was followed in this hypothetical example and the three stakeholder positions ($S_j$) were obvious from the beginning:

$S_1$:  There is a trivial difference in monthly survival probability and this variation cannot be attributed to the application of *Llikmalc*. There is no treatment effect.

$S_2$:  There is a substantial acute survival effect due to the treatment, lasting one month following the aerial application of *Llikmalc*.

$S_3$:  There is a substantial acute survival effect due to the treatment, lasting one month, followed by a month-long chronic survival effect.

TABLE 6.3.  Summary of the evidence for the controversy over the chemical *Llikmalc* and its effect on hen clams.

| Stakeholder position | Log $\mathcal{L}$ | $K$ | $\Delta_i$ | Model probability |
| --- | --- | --- | --- | --- |
| $S_1$ | −6,140.52 | 10 | 19.63 | 0.000 |
| $S_2$ | −6,108.51 | 11 | 5.40 | 0.063 |
| $S_3$ | −6,096.26 | 12 | 0.00 | 0.937 |

The agreed upon protocol made it clear that if a stakeholder's position was unsupported, then the others expected that party to yield. One can see that the stakeholder positions are analogous to the science hypotheses. Here, models must be developed to represent each stakeholder position. Here, an important aspect is that each stakeholder is free to derive their own model to best represent their position. In fact, they might be encouraged to hire expertise in this area. This approach is far different that trying to get all the stakeholders to agree on a single model.

There were many complications in the hen clam study (e.g., overdispersed data, at least 2 different approaches to modeling the recapture probabilities, replicates). I will show only enough of the results to illustrate the type of results one might expect (Table 6.3).

Here it is clear from Table 6.3 that stakeholder position 1 is essentially without support (model probability is 0.000055) and its proponent is expected to yield his position. Most of the support is for stakeholder position 3; the evidence ratio $E_{2,3} = 14.8$ might be viewed as moderate support for the chronic effect. One could examine the estimate of the chronic survival effect and its precision and make further judgments about the importance of chronic effects. The importance of acute effects have been clearly established with moderate evidence concerning a further chronic survival effect. Model averaging could be done to best estimate both acute and chronic survival effects.

The analysis under an information-theoretic approach is the easy part in conflict resolution; it is getting opposing parties to agree on a fair protocol that is often the challenging part. Still, the underlying driving force is the fairness implied in a coin flip, assuming parties that there will be no surprises, and getting them to understand that this is their opportunity to demonstrate to the others that their position is clearly justified and, therefore, will have strong empirical support. There must be a clear statement as to what is expected in the event of various outcomes. I have been part of a large team of people that have used variations of this approach on the northern spotted owl – old growth forest controversy (see Anthony et al. 2006). This has been North America's largest environmental controversy, spanning some 3–4 decades and the data set has been valued at approximately $40M. The protocol has worked well and it is being continually refined. Details of the protocol development and (multimodel) results are given in Anthony et al. (2006).

This approach can be expected to be useful in only a small proportion of existing conflicts. If people are hired to pressure for a particular position against all reason or evidence, this approach will clearly not work and the issue might as well go on to the courts for partial resolution in the long term. Further, if data are not central to the issue, then this approach will not work. However, I think there are many conflicts or controversies within groups of scientists or managers where this approach has potential. Freddy et al. (2004) present a case where the approach might be judged to be useful, but there were difficulties and compromises.

I have seen controversies where some stakeholders withhold their judgment on an ongoing study until the results are in (the coin lands and is inspected). Then, depending on the result, they are either supportive and in agreement or wildly opposed to everything done by the study group. This scenario should be carefully avoided. The use of information-theoretic approaches to aid in the resolution of conflicts is just one application outside the science realm. While this is primarily a primer on science applications, there are a host of other applications in other arenas that are important.

## 6.7    Remarks

Heuristically, $\hat{c}$ adjusts sample size downward in the face of overdispersion (K. Burnham, personal communication). For count data the $\log(\mathcal{L})$ can be written as

$$\sum n_i \log(\pi_i) = n \sum \left( \frac{n_i}{n} \log(\pi_i) \right),$$

where $n = \Sigma n_i$ = sample size. Thus, effective sample size is taken as $n_c = n/\hat{c}$, where $\hat{c} > 1$ and $n_c < n$.

The importance of model selection bias is hard to fathom for a person new to this issue. In some areas of science, I think nearly half of the research work involves the "three demons" – many predictor variables, little science to guide the data collection and modeling, and small sample size. If the results of such work were merely worthless it might not be so bad. However, the results are actually deceiving in that bias suggests the importance of things that are actually not important (i.e., spurious). Ideally, there needs to be a greater awareness of model selection bias and its importance.

There are two cases where "all possible models" finds useful application. First is the ranking of relative variable importance. Second is in computing shrinkage estimates to lessen model selection bias. In both applications, there is a need to put variables or parameters being summed or averaged on an equal footing. In these cases, inferences are not being drawn from the careless, unthinking consideration of "all possible models." Instead, "all possible models" is a device to achieve a proper balance as an intermediate step in a particular analysis type. Beyond these two exceptions, one should not run all the possible models as this is poor practice.

# 7
# Summary

Kei Takeuchi (1933–) was born in Tokyo, Japan, and graduated in 1956 from the University of Tokyo. He received a Ph.D. in economics in 1966 (Kei-zaigaku Hakushi) and his research interests include mathematical statistics, econometrics, global environmental problems, history of civilization, and Japanese economy. He is the author of many books on mathematics, statistics, and the impacts of science and technology on society. His 1976 paper, although obscure and in Japanese, is important as it gives the general result from Kullback–Leibler information, now called TIC in honor of his name. He is currently a professor on the Faculty of International Studies at MeijiGakuin University and Professor Emeritus, University of Tokyo.

I will provide a brief summary of some of the main issues. The remarks below are written from a science perspective because that is what model based inference is about. I wrote this text for others interested in good science strategies and effective methods and the important concept of *evidence*. Application of the information-theoretic approaches are very broad and potentially useful over a very wide range of science and nonscience applications.

## 7.1   The Science Question

The central science question is of critical importance and one must always ask if the question is worthy of study and well focused. The emphasis of this textbook is on a science philosophy that encourages hard thinking to derive a small set of plausible science hypotheses, $H_i$. I think this issue might often take a good person a substantial amount of time and mental effort over several weeks or months. Here one must work hard to define a set of good alternative hypothesis concerning the overall science question. Study of the literature is often a starting place; here one is encouraged to read broadly and not just on the very specific species or process of interest. One should confer with others, attend relevant meetings, use e-mail to correspond with others, ask questions, and try to gain new insights. The emphasis should be on thinking of the various alternatives.

This hard thinking process must go far beyond notions of a null hypothesis. The derivation of a small set of plausible, alternative hypotheses is both difficult and rewarding. This is not something that can be done in an afternoon or a few days; one should be prepared to put their mind to this critical matter. Chamberlin wrote of "…the thoroughness, the completeness, the all-sidedness, and the impartiality of the investigation." He stated, "There is no nobler aspiration of the human intellect that the desire to compass the cause of things." Finally, he believed, "The vitality of the study quickly disappears when the object sought is a mere collection of dead, unmeaning facts." Akaike (Kyoto Award ceremony in 2007) advised, "Select one problem and continue to pursue it until you find the perfect solution."

Before the investigation can move ahead, the alternative hypotheses must be in place. Ideally, data collection (i.e., study design) would be somewhat optimized to try to separate the support for these alternatives and lessen model selection uncertainty. From there one wants to provide measures of quantitative evidence for members of this set and gain a comprehension and understanding of the results. Finally, the set evolves as implausible hypotheses are identified and deleted, remaining hypotheses are refined and strengthened, and new hypotheses are suggested. Some higher dimensioned models with low support might be kept if the sample size is to increase substantially for the next data set. It is this notion of evolving sets that can allow very rapid progress in a field of science. This evolution can provide fast learning if used effectively. This process does not always prevent mistakes or occasionally taking the wrong path; but science has a way of backing up and correcting these setbacks.

I feel there is often far too much descriptive work done in many of the life sciences; this seems particularly true in my fields of ecology and natural resource issues. Some *a priori* thinking can lead to a more confirmatory approach and this has a variety of rewards. We all need to think more about our strategy for doing good science. Issues such as random sampling and scope of inference by defining the population to which inductive inferences are to be made seem so fundamental; I think it is a disservice to continue to accept research papers where such basic things are clearly lacking. A culture needs to be developed to enforce and expect higher standards in our science.

## 7.2   Collection of Relevant Data

The collection of relevant data should deserve special attention. Utmost care should be exercised and this is not the place for volunteers unless carefully trained and closely supervised. A great deal is known about the proper design of experiments and valid sampling protocols. There are dozens of good books on both of these important topics and there is no excuse for collecting data that are fundamentally flawed. Still, I see data collected from convenience sampling where any valid inductive inference from the sample to the population is precluded. In some cases the population of interest is not even defined. I see obvious confounding in experiments and a lax attitude where many variables are measured just because they are easy to measure. Many fields in the life sciences could benefit from more coursework in experimental design and sampling theory. The information-theoretic approaches are not meant to fix bad data, we must accept these challenges as a way to make progress in our science.

Large sample size conveys many advantages in the empirical sciences as does the use of many replicates. Estimators have better performance, precision is enhanced, and evidence for the alternative hypotheses is sharpened; all of these allow better understanding as a result. I see many papers that ask good science questions but they have only 20–50 samples and the need to estimate at least, say, 6–8 parameters. In such cases, there is relatively little information in the data and valid inferences may tend to be shallow and somewhat uninteresting.

## 7.3   Mathematical Models

Information is buried within the data and much of this information can often be extracted by using a mathematical model. Good models of hypotheses are essential in empirical science and are the basis for rigor in the investigation. Soule (1987) suggests, "Models are tools for thinkers, not crutches for the thoughtless." Importantly, the inductive inference is model based. Modeling is both an art and a science and this is a place where consultation with a statistician might be very helpful. Ideally, one hopes that there is a one-to-one correspondence between the $j$th hypothesis and its model.

Modeling is done to evaluate alternative hypotheses, gain insight into model structure, allow predictions, aid in variable selection in regression, and provide objective means of smoothing to identify patterns in the data. Modeling is an essential aspect of empirical science.

## 7.4    Data Analysis

Data analysis begins with the estimation of the unknown parameters and their covariance matrix for each model (these important issues are not the subject of this book; however, Appendix A provides a brief overview of likelihood methods). Other statistics also need to be provided (e.g., adj$R^2$, goodness-of-fit assessments, residual analyses) as these help in the critique of model assumptions. These procedures provide assurances that at least some of the models in the set are worthwhile. Then, one must have the value of the residual sum of squares or the value of the maximized log-likelihood for least squares or likelihood approaches, respectively. These values are the basis for the evidential approaches.

Several things can go awry here: using "all possible models," mixing response variables, counting estimable parameters incorrectly, doing data dredging in the middle of attempting an *a priori* analysis, failure of algorithms to converge (Appendix A.7), etc. Over-fitting and spurious effects should be avoided (see Appendix F). Advice and review by a person in the statistical sciences might be carefully considered at this stage.

## 7.5    Information and Entropy

The ability to quantify information has opened many important doors in science and technology. Boltzmann's entropy is the negative of Kullback-Leibler information and these are fundamental to deep science. Akaike found a link between expected K–L information and the maximized log-likelihood function and this was a pivotal breakthrough. The log-likelihood is a natural estimate of entropy. Akaike's AIC exploited this link and provided an asymptotic correction of bias. A second order bias correction was soon found and this is important to use in applications. While probabilities are multiplicative, information and entropy are additive. These fundamental quantities lead to ways to obtain a formal "strength of evidence" for alternative science hypotheses.

## 7.6    Quantitative Measures of Evidence

I ask graduate students, "what justifies a conclusion." This is a vexing problem for some students as well as professionals in the field. I think an answer relates primarily to "valid methodology." It is the methodology that must be assessed to judge an inference or conclusion: it is the rigor of the *process* that is important.

Hypotheses can be easily ranked using the $\Delta_i$ values. These values are pivotal in various measures of evidence as they are on the scale of information. Being able to quantify information loss is very important in applied science.

Plausible hypotheses exist only within a fairly narrow band; perhaps 0–8 or 12 on a scale of information loss if the independence assumption can be met.

It is simple to obtain the (relative) likelihood of each model $i$, given the data: $\mathcal{L}(g_i|x)$. These are useful measures of the strength of evidence for science hypotheses and do not depend on other models in or out of the set.

It is equally simple to compute the discrete probability of each model $i$, given the data: $\text{Prob}\{g_i|x\}$. These measures of strength of evidence are conditional on the set of hypotheses and their models.

Finally, an *evidence ratio* is just the quotient of 2 model probabilities (or 2 model likelihoods) and is another way to effectively quantify the evidence for any two hypotheses, as represented by their models. Only the two models being compared are relevant here, regardless of other models in or out of the set.

The hard science stops with the provision of various quantitative measures of the evidence. Following this, value judgments can be made to qualify the evidence. The investigator is certainly able to make their value judgments as are others. In many cases, honest observers will reach the same qualitative conclusions about the strength of evidence, while in other cases there may be honest differences in this interpretation.

This distinction helps scientists with the contentious issue of "advocacy." Scientists certainly have the right to clearly state and stand behind the objective, quantitative result; this is not advocacy. The qualification of the result can sometimes push the issue into an advocacy position – these become value judgments.

## 7.7   Inferences

Most inference methods in the life sciences are inductive and statistical. Properly done, both allow rigorous inference from the sample data to the population sampled. Initially, there is interest in the estimates of model parameters and measures of the uncertainty but one must determine which model to use as a basis for these estimates.

It now seems clear that final inferences should routinely be based on all the models in the set – multimodel inference. This important extension allows information in the data from models other than the best to be used in making inferences. The main tools here at the moment are model averaging and the use of unconditional variances to incorporate model selection uncertainty into estimates of precision. Both approaches are easy to compute and to understand. While there may be cases where inference is sensibly confined to a single model; however, the use of all the models will be commonplace. Multimodel inference is most often the effective path to reliable evidence.

There needs to be increased awareness of conditions that lead to spurious effects (e.g., Freedman's paradox). If one has little background science to guide the hypothesizing and modeling, small sample size, many predictor variables, and many models, the results will likely be largely spurious. People often fear they will miss an effect that is contained in the data; however, they should have an equal fear that they will find something that is not there at all (i.e., spurious). A type of model averaging is useful in reducing the important issue of model selection bias.

## 7.8   *Post Hoc* Issues

I encourage some *post hoc* examination of the data. This might include the addition of new hypotheses and models to represent them or slight changes to several of the better models. Such examination can include residual analyses and goodness-of-fit results leading to additional models. Because such examination and subsequent modeling are based on the same data, the conclusions from such activities must be recognized as being weaker than the more confirmatory inferences.

## 7.9   Final Comment

Given some background science and philosophy (Chaps. 1 and 2), it can be helpful to view the information-theoretic approaches at three different levels. The first level is conceptual and entails the Principle of Parsimony and Occam's razor (Chap. 2). The second level is the rigorous target of model selection – expected Kullback-Leibler information (Chap. 3). The third level provides a simple approach to application – various forms of Akaike's information criterion and quantitative measures of strength of evidence (Chaps. 3 and 4). These approaches are simple to compute and seem compelling. The entire approach seems to encourage people to be good scientists and allow fast learning and understanding.

The cutting edge in model based empirical science is the concept of multimodel inference (Chap. 5). There are substantial advantages to be realized in basing inferences on all the models in the set. Doing so is computationally trivial and easy to understand and interpret.

# Appendices

## Appendix A: Likelihood Theory

Likelihood methods are much more general, far less taught in applied statistics courses, and easier to understand as a concept or procedure than least squares. The material in much of this book relies on an understanding of likelihood theory, and so a very brief introduction is given here. While likelihood methods underlie both frequentist and Bayesian statistics, there are no more than a handful of applied books on this important subject (examples include McCullagh and Nelder 1989; Edwards 1992; Azzalini 1996; Morgan 2000; Severini 2000; Pawitan 2001) and none of these constitute easy reading.

### A.1 Likelihood Functions

The first key point is that the likelihood function is a product of probabilities. The concept can be illustrated by considering events (or outcomes) that can be observed (e.g., the number of "heads" observed from flipping a coin $n$ times). The set of these observations constitute the data. Specifically, the data from a coin flipping study are the number of heads ($y$) and the number of tails ($n-y$) from $n$ coin flips. The *probability* of such events can be "assigned." Underlying each time a head is observed is the probability of a head; call this $p$. Underlying each observation of a "tail" is it's probability; call this $1-p$.

Tacit assumptions have been made; these are often termed "independent and identically distributed, *iid*. It is easy to believe that the outcomes of coin flips are independent. Whether a single coin is flipped $n$ times or $n$ coins are flipped once, surely one outcome does not influence the next. The term *identically distributed* relates to each coin having the same properties; if one coin has the probability of a head as some value $p$, the others have that same value (this condition is also known as parameter homogeneity). These are important assumptions; for example, unless independence is assumed probabilities are not simply multi-plied and log-likelihoods cannot be summed. Both of these *iid* assumptions can fail in many applications in the life sciences (see Sect. 6.2).

Note, the sum of the number of heads and tails $= n = y + (n-y)$. Likewise, the sum of the probabilities is $1 = p + (1-p)$. All the events and their probabilities must be accounted for under basic rules of probability. Assume a coin is flipped 11 times ($n = 11$) and 7 heads ($y$) are observed. Then the likelihood function ($\mathcal{L}$) for this could be written as the product (note the order is not important),

$$\mathcal{L} \propto pppppp(1-p)(1-p)(1-p)(1-p).$$

Some simple notation allows this to be written in a more useful form (where $\propto$ means "proportional to")

$$\mathcal{L} \propto p^y(1-p)^{n-y}$$

or for example

$$\mathcal{L} \propto p^7(1-p)^4.$$

Now it should be clear that this is the binomial model. The above shows the likelihood function is proportional; its exact form must include the binomial coefficient

$$\mathcal{L} = \binom{n}{y} p^y (1-p)^{n-y},$$

where

$$\binom{n}{y} = \frac{n!}{y!(n-y)!}$$

and is read "$n$ choose $y$" and is the number of ways a sample of size $y$ can be selected from a population of size $n$. The binomial coefficient $\binom{n}{y}$ does not contain the unknown parameter $p$ and is often omitted for estimation of model parameters. The key here is to focus on the fact that likelihood functions are the product of probabilities. These probabilities come from assigning underlying probabilities to observed events (the data). This is formalized as

$$\mathcal{L}(p \mid n \text{ and } y, \text{binomial}) = \binom{n}{y} p^y (1-p)^{n-y},$$

and is read "the likelihood of the unknown parameter $p$, given the data ($n$ and $y$) and the model (binomial). The likelihood function allows the estimation of unknown parameters, given the data and the model ($g$). The scientist has data and can assign probabilities underlying the data if the model is given (or can be selected). This paves the way for a way to estimate the value of parameters in the model.

There are important distinctions between the terms probability and likeli-hood. Likelihood is relative or comparative; likelihood values do not sum or integrate to 1. Likelihoods are not probabilities. Likelihood values are like raffle tickets. If you have 14 tickets and Barney has only one ticket then the likelihood of you winning the raffle, relative to Barney winning, is 14:1. Likelihoods are functions of the unknown parameters ($\theta$), given the data ($x$) and the model ($g$); $\mathcal{L}(\theta|x, g)$. In contrast, probabilities sum or integrate to one and are absolute. Probability functions and distributions are functions of the data, given the value of the parameters and a model; $p(x|\theta, g)$. Both probabilities and likelihoods are conditional on various things. Both quantities are useful in inductive inference, but they are different (even though lay people might use these interchangeably).

Clearly, the likelihood is a function of (only) the unknown parameter ($p$ in this example), given the model upon which $\mathcal{L}$ is based. Those familiar with the binomial probability model will note the similarity with the binomial likelihood. The probability model of the data and the likelihood function of the parameter are closely related; they merely reverse the roles of the data and the parameters, given a model. The important point to remember is that the likelihood function is always a product of the probabilities.

Thus, given the data ($y$ and $n$) and the binomial model, one can compute the *likelihood* that $p$ is 0.15 or 0.73 or any other value between 0 and 1. The likelihood (a relative, not absolute value) is a function of only the unknown parameter $p$. Given this formalism, one might compute the likelihood of many values of the unknown parameter $p$. The likelihood of 4 values of $p$ are tabulated below.

| $P$ | $\mathcal{L}$ |
|-----|-----|
| 0.3 | 0.0173 |
| 0.5 | 0.1611 |
| 0.7 | 0.2201 |
| 0.8 | 0.1107 |

Clearly, some values of $p$ are much more *likely* than others and this is invariant to any scaling of the data. In fact, $p = 0.7$ is 12.7 (= 0.2201/0.0173) times more likely than the value of $p = 0.3$. Given the ability to compute the likelihood of various values of $p$, Fisher reasoned that the best estimate of the unknown parameter $p$ would be the one that was "most likely." Hence the term *maximum likelihood estimate* or MLE. In the values tabulated above, $p = 0.7$ is the most likely. If the derivative of the analytical form of the likelihood were used to compute the exact maximum of the entire function, we would see that the MLE is 0.63636. This estimate could also be gotten using numerical methods and that is what is done in practice with real problems.

It seems compelling to pick the value of the unknown $p$ that is "most likely." Likelihood theory includes asymptotically optimal methods for estimation of unknown parameters and their variance–covariance matrix, derivation of hypothesis tests, the basis for profile likelihood intervals, and other important quantities. Likelihood is the backbone of statistical theory, whereas least squares can be viewed as a limited, special case (but certainly an important case).

## A.2   Log-Likelihood Functions

For many purposes the natural logarithm of the likelihood function is essential; written as $\log(\mathcal{L}(\theta|\text{data}, \text{model}))$, or $\log(\mathcal{L}(\theta|y, \text{model}))$, or if the context is clear, just $\log(\mathcal{L}(\theta))$ or even just $\log(\mathcal{L})$. Thus, taking logarithms

$$\log(\mathcal{L}(\theta \mid y, \text{model})) = \log\binom{n}{y} + y \cdot \log(p) + (n - y) \cdot \log(1 - p).$$

Often, one sees notation such as $\log(\mathcal{L}(\theta|y))$, without making it clear that a particular model is assumed. An advanced feature of $\log(\mathcal{L})$ is that it, by itself, is a type of *information* concerning the unknown parameters ($\theta$) and the model. A property of logarithms for values between 0 and 1 is that they lie in the negative quadrant; thus, values of the log-likelihood function are negative (unless some additive constants have been omitted). Figure A.1 shows a

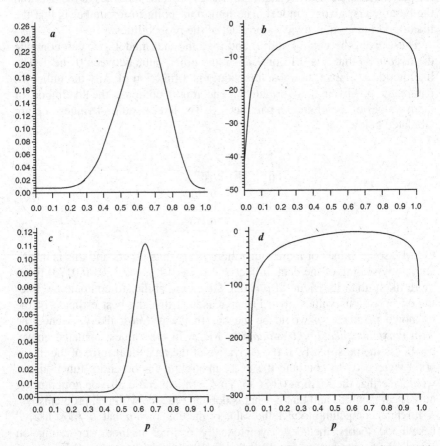

FIG. A.1.   Plots of the binomial likelihood (*a*) and log-likelihood (*b*) function, given $n = 11$ penny flips and the observation that $y = 7$ of these were heads. Also shown are plots of the binomial likelihood (*c*) and log-likelihood (*d*) function, given a sample size 10 times larger; $n = 110$ penny flips and the observation that $y = 70$ of these were heads.

plot of the likelihood (*a*) and log-likelihood (*b*) functions where 11 coins are flipped, 7 heads are observed, and the binomial model is assumed. The value of $p = 0.6363$ maximizes both the likelihood and the log-likelihood function; this value is denoted as $\hat{p}$ and is the maximum likelihood estimate (MLE). Relatively little information is contained in such a small sample size ($n = 11$) and this is reflected in the broad shape of the plots. Had the sample size been 10 times larger, with $n = 110$ and 70 heads observed, the likelihood and log-likelihood functions would be more peaked (Fig. A.1). In fact, the sampling variance is derived from the shape of the log-likelihood function around its maximum point. Finally, the value of the log-likelihood function at its maximum point is a very important quantity and it is this point that defines the maximum likelihood estimate. In the example with 11 flips and 7 heads, the value of the maximized log-likelihood is $-1.411$ (Fig. A.1b). Thus, when one sees reference to a maximized $\log(\mathcal{L}(\theta))$, this merely represents a numerical value (e.g., $-1.411$). The value $-1.411$ is computed using the binomial coefficient

$$\binom{11}{7} = \frac{11!}{7!(11-7)!} = 330.$$

Specifically, the value of the maximized log-likelihood function is

$$\log(\mathcal{L}\ (p\,|\,x, \text{binomial})) = \log\binom{n}{y} + y\cdot\log(p) + (n-y)\cdot\log(1-p),$$

$$\log(\mathcal{L}\ (p\,|\,7,11, \text{binomial})) = \log\binom{11}{7} + 7\cdot\log(0.6363) + (4)\cdot\log(1-0.6363)$$

$$= 5.799 + 7(-0.452) + 4(-1.012)$$

$$= -1.411.$$

The value of the log-likelihood function $\log(\mathcal{L}) = -1.411$. Then, AIC $= -2\log(\mathcal{L}) + 2K$ is simply $-2(-1.411) + 2(1) = 4.822$. Software for computing MLEs always give the value of the maximized log-likelihood or the deviance (which is $-2\log(\mathcal{L})$ and is the first term in AIC and AICc). Thus, computation of AICc is trivial once the MLEs have been found.

Those using LS to get estimates in linear models can easily compute the value of the maximized log-likelihood function by the simple mapping

$$\log\big(\mathcal{L}(\hat{\underline{\theta}})\big) \sim -\tfrac{1}{2}n\log(\hat{\sigma}^2),$$

where $\hat{\sigma}^2 = \text{RSS}/n$ (the MLE). This result is important in model selection theory as it allows a simple mapping from LS analysis results (e.g., the RSS or the MLE of $\hat{\sigma}^2$) into the maximized value of the log-likelihood function for comparisons over such linear models with normal residuals. Note that the log-likelihood is defined up to an arbitrary, additive constant in this usual case. If the model set includes linear and nonlinear models or if the residual distributions underlying the models differ (e.g., normal, gamma, and log-normal), then all the terms in the log-likelihood must be retained, without omitting any constants. All uses of the log-likelihood are relative to its maximum, or to other likelihoods at their maximum, or to curvature of the log-likelihood function at the maximum.

The variance–covariance matrix can be found from the log-likelihood function; this is a more technical subject and I will only provide a glimpse into Fisher's approach. The variance is directly related to the shape (peakedness) of the log-likelihood function near the maximum point. The more peaked the smaller the variance and vice versa. If there are 3 unknown parameters, then the variance–covariance is a square matrix with dimension 3. The 3 variances appear on the diagonal, while the covariances appear in the off-diagonal elements. [Elements of this matrix come from second mixed partial derivatives of the log-likelihood function with respect to the parameters. This is a very general and useful procedure, but often seems difficult when first encountered; we will not take this issue further here.]

The likelihood function $\mathcal{L}(\theta|x$, model) makes it clear that for inference about $\theta$ the data and the model are taken as *given*. Before one can compute the likelihood that $\theta = 0.53$, one must have data and a particular statistical model. While an investigator will have empirical data for analysis, it is unusual that the model is known or given. Rather, a number of alternative model forms must be considered as well as the specific explanatory variables to be used in modeling a response variable. This issue includes the *variable selection problem* in multiple regression analysis. If one has data and a model, LS or ML theory can be used to estimate the unknown parameters ($\theta$) and other quantities useful in making statistical inferences.

Model selection relates to fitted models; given the data and the form of the model, then the MLEs of the model parameters have been found ("fitted").

## A.3   Why Likelihood Theory?

The review above has been in terms of only one model (the binomial) with a single unknown parameter, but the principles extend to other models and models with hundreds of unknown parameters. The theory is worth the effort to learn and be comfortable with. Reasons for this include

- Likelihood and log-likelihood functions form the general basis for deriving estimates of unknown parameters in the models of science hypotheses and their variance–covariance matrix as measures of precision
- Log-likelihood functions are the basis for profile likelihood intervals. These allow for asymmetric intervals and avoid the notion of repeated sampling and the awkward definition of the usual frequentist intervals
- Likelihood and log-likelihood values are the basis for hypothesis tests – the likelihood ratio tests (LRT) and goodness-of-fit tests in particular (however, these are of little use in model building or model selection)
- Model selection based on Kullback-Leibler information

## A.4    Properties of Maximum Likelihood Estimators

MLEs are asymptotically optimal; that is, as sample size gets "large" they enjoy the following important properties:

- Normally distributed
- Minimum variance
- Unbiased

In addition, linear or nonlinear transformations of an MLE to estimate another parameter are also MLE. For example, mean life span $\bar{L}$ is defined as $1/\log(S)$. An estimator of mean life span is

$$\hat{\bar{L}} = 1/\log(\hat{S}),$$

where $\hat{S}$ in an MLE. This being the case, then one can say that $\hat{\bar{L}}$ is also MLE. This is a very important property in application.

## A.5    Deviance

A useful quantity in likelihood-based inference is the deviance,

$$\text{Deviance} = -2\log(\mathcal{L}(\hat{\theta} \mid x,g)) + 2\log(\mathcal{L}_s(\hat{\theta} \mid x, g)),$$

where $\mathcal{L}$ is a "saturated" model. In model selection, this $\mathcal{L}_s$ term is constant across models and can usually be omitted. In other situations the saturated model would produce $\log(\mathcal{L}_s) = 0$; hence, there is a basis to say deviance $= -2\log(\mathcal{L}(\hat{\theta} \mid x, g))$. Thus, for the issues here, deviance $= -2\log(\mathcal{L}(\hat{\theta} \mid x, g))$ and is a measure of lack of fit and is the first term in AIC and AICc.

## A.6    Likelihood Ratio Tests

Likelihood ratio tests (LRT) can be used to compare two nested models; the form of the test is suggested by its name

$$T = -2\log\left(\frac{\mathcal{L}_s(\hat{\theta} \mid x,g)}{\mathcal{L}_g(\hat{\theta} \mid x,g)}\right),$$

where the simpler model (s) has fewer parameters than the general model (g) – seen as subscripts. [Note the appearance of the $-2$ again.]
Asymptotically, the test statistic ($T$) is distributed as a chi-squared variable with degrees of freedom equal to the difference in the number of parameters between the two nested models. LRTs can also be expressed in terms of the difference between the two deviances.

## A.7   A Likelihood Version of $R^2$

Nagelkerkle (1991) provided a near analog to the $R^2$ of least squares, we will denote this as $\mathcal{R}^2$. Let $\ell(\hat{\theta})$ and $\ell(0)$ denote the maximized log-likelihoods for the fitted model of interest and the null model, respectively. Start with

$$R^2 = 1 - \exp\left\{-\frac{2}{n}(\ell(\hat{\theta}) - \ell(0))\right\}$$

and then rescale to allow a maximum of 1 by defining

$$\max R^2 = 1 - \exp\left\{\frac{2}{n} \cdot \ell(0)\right\}.$$

and finally the rescaled value we want

$$\mathcal{R}^2 = R^2 / \max R^2.$$

Often, the statistic $\mathcal{R}^2$ is optimistic and it is not an exact analog to the usual $R^2$ in linear models. Still, this approach is useful and easy to compute. In addition, other approaches have been developed such as the "analysis of deviance," which is closely allied with the usual $R^2$ in regression.

## A.8   Potential Problems

Virtually all applications of likelihood methods for real problems are done numerically. That is, calculus is not used to find the maximum of multidimensional functions; instead, sophisticated numerical routines have been found years ago to perform these tasks.

The first problem is that the function, at least in one dimension, is very flat and the numerical routine cannot identify the "exact" maximum point. There are several reasons that might cause this; however, the software usually outputs a message that it failed to converge. The user might restart the routine using the provisional values of $\hat{\theta}$ available when the routine last stopped. Alternatively, one might start over using a different starting value for $\hat{\theta}$.

The second problem is that a log-likelihood function might have multiple local maxima (modes) and one must worry that the numerical routine will find a suboptimal maximum point and this is unknown to the user. Here, one might try different starting values or use some other numerical routine (e.g., simulated annealing). Most of the commonly used statistical distributions are in the so-called "exponential family" and these carry a guarantee of unimodality (however, mixture distributions of these common forms do not).

# Appendix B: Expected Values

Statistical expectations of estimators or other expressions are often useful in a variety of ways. Such expectations can be thought of as an "average" taken over all possible samples of size $n$ (see Wackerly and Mendenhall 1996). This process is simple when working with discrete random variables. The expectation of a discrete random variable $x$ is defined as

$$\mathbf{E}(x) = \sum_i x_i p(x_i),$$

where $p$ is the probability of being in class i. Consider a population of size $N = 4$ and a sample of size 2. The binomial coefficient $\binom{N}{n}$ is read "N chose n" or, in this example, $\binom{4}{2}$ is "4 chose 2" = $4!/[2! \times (4-2)!] = 6$. This is an effective way to compute the number of ways a sample of size 2 can be drawn from a population of size 4. In general, the binomial coefficient is

$$\binom{N}{n} = \frac{N!}{n! \times (N-n)!},$$

where ! means factorial. Let $N = 5$, then 5! is $5 \times 4 \times 3 \times 2 \times 1 = 120$.

Now consider a population of 4 rats (rat A, B, C, and D) each with a number of ticks. An exact count of the number of ticks on each rat has been made; rat A has 2 ticks, rat B has 4 ticks, rat C has 2 ticks, and rat D has 8 ticks. As we have an exact count of the number of ticks on all the rats in the population, we can compute the mean number of ticks per rat as a population parameter; denote this parameter as $\mu$. The value of $\mu$ in this simple example is merely the total number of ticks ($2 + 4 + 2 + 8 = 16$) divided by the number of rats (4). This gives the parameter as $\mu$ = an average of 4 ticks per rat. So, the population parameter in this example is known, $\mu = 4$.

We must now summarize all possible samples of size 2 that could be drawn from the population of size 4; we know from the binomial coefficient that there are 6 such samples of size 2 possible. The sample data are summarized below:

| Sample, $i$ | No. ticks | Sample mean | $\hat{\mu}$ |
|---|---|---|---|
| 1 | AB | 6 | 3 |
| 2 | AC | 4 | 2 |
| 3 | AD | 10 | 5 |
| 4 | BC | 6 | 3 |
| 5 | BD | 12 | 6 |
| 6 | CD | 10 | 5 |

Each of the 6 sample means $\hat{\mu}_i$ is a maximum likelihood estimate. The expected value of the MLE $\hat{\mu}$ is written as $E(\hat{\mu})$ and is the average of the 6 sample means

$$(3+2+5+3+6+5)/6 = 4.$$

Thus, $E(\hat{\mu}) = 4$. This is the average of all possible samples from the population of size 4 for samples of size 2. The notation "$E(\cdot)$" is an operator meaning "take the expectation of $(\cdot)$." One reason for taking statistical expectations is in assessing the bias of an estimator. Bias is also an average quantity and defined as

$$\text{Bias} = E(\hat{\theta}) - \theta$$

where $\theta$ is some parameter of interest. In the rat example, bias = $E(\hat{\mu}) - \mu = 4 - 4 = 0$, or unbiased. Expectations of continuous random variables also exist; integrals replace summation operators, but the principle remains the same.

A second type of expectation is useful in parameterizing some types of models. Consider a sample of size $R_2$ sea turtles marked and released in year 2 of a conservation biology study. Four years after release, $r_{25}$ turtles are killed (as bycatch) in a primitive fishery and reported to conservation authorities. The notation $r_{25}$ reflects the number of turtles recovered dead in year 5 from those marked and released in year 2. So, under a model that allows survival and reporting probabilities to vary by year, we can write down the expectation of $r_{25}$, i.e., $E(r_{25})$. Here the expectation operator ($E$) asks for the analytical expression of the count $r_{25}$, given a model. We note that to have been killed and reported in year 5, the turtles had to survive the yearly intervals 2–3, 3–4, 4–5, they died in year 5, and were reported in year 5. Thus, under the time-specific model

$$E(r_{25}) = R_2 S_2 S_3 S_4 (1 - S_5)\lambda_5,$$

where S is the annual survival probability in year $j$ and $\lambda$ is the annual reporting probability in year $j$. In this case, one would like estimates of the 5 model parameters and their sampling covariance matrix using maximum likelihood methods. The expectation changes if a different model is hypothesized where the parameters are nearly constant across years (an approximation as we know that there is some variation in the parameters across years). Here

$$E(r_{25}) = R_2 SSS(1 - S)\lambda = R_2 S^3 (1 - S)\lambda.$$

Under this model there are only 2 parameters, $S$ and $\lambda$. The expectation operator is used often in statistics.

A final example is the expectation of an encounter history matrix used in capture–recapture and occupancy models. For each sampling occasion $i$ let "1" denote encountered and "0" denote not encountered. As an example, take the encounter history for manatee no. 17 over 8 sampling occasions:

$$\{11001101\}.$$

The "1" in the final column (representing year 8) makes it clear that the animal was still alive in the 7th (last) year. Thus, the expectation must contain 7 annual survival probabilities, $\phi_1, \phi_2, \ldots, \phi_7$, related to the 7 intervals defined by the 8 occasions (this reasoning assumes the model has year-specific parameters). These models condition on the first occasion and so there is no encounter probability (denoted as $p_1$) for occasion 1. Note, this manatee was encountered on occasion 2, 5, 6, and 8, following its initial capture. Thus, the expectation must contain $p_2$, $p_5$, $p_6$, and $p_8$. Finally, this animal was not encountered on occasions 3, 4, and 7 and so the expectation must include $(1-p_3)$, $(1-p_4)$ and $(1-p_7)$. In summary

$$E\{11001101\} = \phi_1 \phi_2 \cdots \phi_7 p_2 p_5 p_6 p_8 (1 - p_3)(1 - p_4)(1 - p_7);$$

however, the order of the parameters is arbitrary. This component of the model has 14 unknown parameters.

As above, the expectation would be different if a different model were hypothesized. For example, if one hypothesized a fairly constant environment and relatively constant sampling effort, then a model with only an average annual survival and encounter probability would yield the following expectation for the same encounter history

$$E\{11001101\} = \phi^7 p^4 (1 - p)^3.$$

This model has only 2 parameters and these parameters and their covariance matrix can be estimated using maximum likelihood methods, given data. Given a specific data set, which of these 2 models is "better"? This is a model selection problem and its solution must take into account the concept of parsimony.

# Appendix C: Null Hypothesis Testing

The central inferential issues in science are twofold. First, scientists are fundamentally interested in estimates of the magnitude of parameters or functions of parameters and their precision: are the effects trivial, small, medium, large, or extra large? Are these effects biologically meaningful or interesting? This is an *estimation* problem whether the data arise from a strict experiment or an observational study. Second, one often needs to know if the effects are large enough, given the data, to justify inclusion in a model to be used for further inference (such as prediction). This is a *model selection* problem and involves the principle of parsimony. These issues are not strongly associated with null hypothesis testing, $P$-values, and rulings about "statistical significance." Null hypothesis testing in the statistical sciences is like protoplasm in biology; they both served an early purpose but are no longer very useful.

Some people still believe that statistics and statistical science are mostly about testing null hypotheses without realizing the uninteresting or trival nature of most such hypotheses. Many null hypotheses are merely strawmen to be struck down and rejected, but little understanding is gained by doing so. We need to move on from the traditional testing approach because it is so uninformative.

Given that many of us were trained in null hypothesis testing, it is easy to cling to the incorrect notion that $P$-values represent a strength of evidence. Royall (1997), Vieland and Hodge (1998:285), and Johnson (1999) provide convincing proof that this is not the case (the reasons are technical in that $P$-values are dependent upon the sample space of both observed and unobserved data). One unsettling issue (there are many) is assigning probabilities to events that were never observed. I urge people to think hard about the differences in approach as illustrated by the European dipper example in Sect. 4.8.

Some authors still see a use for null hypothesis testing when the evidence against this seems, to me, so overwhelming (e.g., Stephens et al. 2005; Steidl 2007); I do not mean to criticize, only to note the large variance component here. I believe that null hypothesis testing will continue to decline as it is replaced by the substantially more relevant methods based on information theory and Bayes' theorem.

# Appendix D: Bayesian Approaches

This appendix assumes the reader has a basic understanding of the Bayesian paradigm. Bayesian approaches have seen tremendous growth and recognition in the past 2–3 decades (Gelman et al. (2003) lists nearly 600 references). This change has been the result of huge increases in computing power and the discovery of powerful numerical methods (i.e., Markov Chain Monte Carlo methods, MCMC, see Chen et al. (2000) and Givens and Hoeting (2005)). Bayesian methods are particularly powerful in coping with a wide class of random effects models (see Sect. 6.5) and will continue to see heavy use in this area. There are many excellent books on Bayesian methods including Carlin and Louis (2001) and Gelman et al. (2003).

Bayesian methods have met with controversy over the past 2.5 centuries; this stems primarily from the subjective nature of early Bayesian approaches. Change has emerged in the thinking of many Bayesians because of the use of "vague" priors; also termed uninformative, colorless, or flat priors. Here the goal is to attempt to withhold any subjective (or "personal") information; thus, the resulting analysis is objective and the parameter estimates are often virtually identical to the MLEs. This change in approach has greatly lessened the strong objection to Bayesian methods in science where subjectivity is to be minimized, not invited or enhanced. Subjective priors on parameters often have utility in nonscience issues; but such priors have been largely rejected in scientific work. Having said that, I must note that the data "swamp" the prior in some science applications and, if this is clearly demonstrated to be the case, then there are no objections with this approach in scientific work. The use of vague priors on model parameters has been a major step forward for the acceptance of Bayesian approaches by scientists.

Bayesian approaches to model selection include the Bayesian Information Criterion (BIC), the deviance information criterion (DIC), and a reversible jump Markov Chain Monte Carlo approach (RJMCMC). DIC is a Bayesian approach but with AICc-like properties and has seen heavy use in the free software WINBUGS and more generally. DIC seems to be the workhorse for Bayesian model selection; however, other approaches also see application.

Bayesian prior probabilities on models are required when dealing with several models. BIC (see Appendix E) has both a Bayesian derivation and a "frequentist" derivation, whereas AIC also has both a Bayesian and "frequentist" derivation. Thus, debate should not be just "Bayesian vs. non-Bayesian" (see Link and Barker 2006); the issues are more substantive than this. Turning beliefs about models into probability distributions has been difficult. Still, I think a goal in Bayesian analysis would be to have the model priors swamped by the data.

The level of education and experience needed to thoughtfully use Bayesian methods is fairly high. One must have a decent background in probability, mathematical statistics, numerical analysis, and programming ($R$ being especially useful) in addition to the subject matter science. This is asking a lot. I encourage research people in the life sciences to seek a PhD level statistician with expertise in Bayesian theory and computation and work collaboratively with them.

Programs such as WINBUGS are useful for smaller problems and can be surprisingly useful for many research problems. Otherwise, the researcher must often write and debug code for the MCMC or RJMCMC algorithms and this can be quite challenging. One must anticipate substantial computer run times as well as programming and debugging issues. The recent text by Givens and Hoeting (2005) provides a review of these issues.

I have a high regard for Bayesian approaches and I expect to see their increasing use in the future. In multilevel random effects models, there is little choice of method and the nature of the MCMC algorithm makes Bayesian approaches a natural for coping with random effects (however, the concept of h-likelihood might provide an alternative at some point). I think more work needs to be done to explore the mutualities between extended likelihood theory and Bayesian methods. Ken Burnham has shown several areas of commonality between what might be called likelihoodists and Bayesians (Burnham and Anderson 2004). Other investigators have found similar convergence and I view these as constructive.

# Appendix E: The Bayesian Information Criterion

Akaike's AIC started one of Claude Shannon's "bandwagons," the first and best known is BIC, the Bayesian information criterion (also called SIC after its founder, Schwarz (1978)). BIC is superficially similar to AIC

$$\text{BIC} = -2\log(\mathcal{L}(\hat{\underline{\theta}})) + K\log(n)$$

but with a different "penalty" term. If $n$ = about 8, then BIC = AIC. In the realistic cases where $n > 8$, the penalty in BIC is slightly larger and there is a tendency for it to select smaller dimensioned models than AIC. Comparisons between BIC and AICc are harder to generalize.

BIC has nothing linking it to information theory, a misnomer. Many Bayesians do not like BIC (e.g., Link and Barker 2006); however, it is not uncommon to see its output by various statistical software packages, thus I will offer a few comments and a comparison. Almost any short summary as to what BIC is supposed to do is probably somewhat wrong or incomplete (including this one). There are a large number of papers about BIC; useful (but inconsistent) summaries can be found in Weakliem (2004). McQuarrie and Tsai (1998) provide the results of elaborate MC simulation studies that include BIC as one criterion. BIC has been rediscovered many times and several elaborations have been published over the years. I will not attempt a thorough review; instead I will offer some overview comments on this issue.

## E.1    Schwarz' Criterion

Schwarz' derivation of BIC does not assume that a true model exists; however, the general setting is that a true model exists, this model is in the candidate set, and the investigator does not know which model is the true one, thus a model selection problem – "find the true model." Schwarz derived the criterion using vague priors on all the model parameters and uniform priors ($1/R$) on models. Bozdogan (1987) termed what would eventually become a class of such criteria, "dimension consistent."

Consistency is a statistical property in estimation theory indicating an estimator with both bias and variance going asymptotically to zero. Consistency has often been touted as BIC's virtue; however, this has no meaning without the false concept of a true model being in the candidate set.

## E.2    Real World Properties

The real issue, then, concerns the properties of BIC when the true model is not in the set and when sample size is less than very large. Such properties are difficult to state clearly as they depend substantially on the nature of the underlying reality. I will outline two extremes, (a) are there only 3–4 large effects (and no other effects) in the underlying process? or (b) are there a wide range (say, 25–80 – if not hundreds) of tapering effect sizes in the underlying process of interest? Some useful generalizations can be given for these cases.

In (a) BIC will often do well in terms of selecting the model with these few and large effects even if sample size is small to moderate (so will AICc). Nearly all MC simulation studies generate data from a model with a few (2–5) large effects (but see McQuarrie and Tsai 1998); thus, the result would seem to show that BIC selects the true model a high percentage of the time.

In (b) BIC will perform poorly in identifying the full extent of reality unless sample size is very, very large. BIC approaches the true model from the left; thus, if sample size is too small, an underfitted model (as judged by full reality) will be selected. BIC will do poorly at selecting the model of complex reality in case (b), unless one has samples sizes in the (I am guessing) millions. Understanding the underlying realities gives little place for BIC to contribute. Link and Barker (2006) offer additional points.

Burnham and Anderson (2002) suggested the notion of a quasi-true model to help with an understanding of BIC's performance in realistic situations; however, even this notion is strained, but at least it points to the target model for BIC selection when a true model is not in the set. BIC does not guarantee a good parsimonious model, or minimum MSE, good confidence interval coverage, or other performance properties.

## E.3   High Probability Assigned to Models that Do Not Fit

BIC has a tendency to give high weight to models that do not fit, as judged by a usual goodness-of-fit test (Burnham and Anderson 2004:293–297). One might hope that if the global model fits, the selected model would also fit: AICc has this property. Under tapering effect sizes and using $\alpha = 0.05$, they found that BIC selected nonfitting models 11.5% of the time with sample size = 50, 15.9% of the time with sample size = 100, and 28.1% of the time with sample size of 500. As sample size increases, the probability of selecting a nonfitting model increases! These results would seem to be disturbing and more work on this issue is warranted.

Reschenhofer (1996) noted that AICc and BIC have very different objectives and target models and should not be directly compared. AICc depends on the given sample size and selects the *fitted* model that minimizes estimated, expected K–L information as the approximating model of full reality. AICc is about approximation and prediction and its target model changes with changes in sample size. Thus, as sample size gets larger, additional effects can be uncovered; this includes reality where there are countless tapering effects. AICc is about "best" models in the sense of approximations to truth and out-of-sample prediction, given the sample size.

## E.4    Predictive Mean-Squared Error

Almost no MC simulation studies have been reported in the literature where data were generated from a model with reasonable complexity (say, a non-linear model with 40–50 or 100 parameters, many correlated covariates, several higher order interactions). Then, over a range of sample sizes, evaluate various selection criteria on predictive mean-squared error (PMSE) or achieved confidence interval coverage for predictions. Burnham and Anderson (2002:300) present the results of a reanalysis of the human body fat data from Johnson (1996). This is a linear regression to predict body fat using 13 predictor variables (= 8,191 models). They took the global model, its MLEs, and covariance matrix and used it as a generating model to simulate 10,000 reps each with sample size 252. I will not give details here except to tabulate some PMSEs ($\times 10^6$) for (a) model averaging (multimodel inference, Chap. 5) used or (b) inference from (only) the best model.

| Method | Model Averaged | Best Model |
|--------|----------------|------------|
| AICc   | 4.8534         | 5.6849     |
| BIC    | 5.8819         | 7.6590     |

AICc has a substantially better PMSE, but note that BIC benefited relatively more from model averaging. More simulation studies to mimic real world phenomenon would be helpful. In these cases, the evaluation should be focused on PMSE instead of the usual "how often does this criterion select the true model"? Of course, the generating ("true") model should not be in the set, a mistake so often seen in the literature.

In summary I would not use BIC unless I was trying to select the generating model from MC simulation. There, a true (generating) model exists and I know if it is in the set. Then, if the generating model mimicked some complex reality and if sample size is very large (e.g., perhaps hundreds of thousands or millions), I would use BIC. Alternatively, if I knew the underlying process had 3–5 large effects (and no smaller effects) I might use BIC even if sample size was modest – this is BIC's element. Putting this in perspective, I would still use BIC in regression settings over step-up, or step-down or stepwise methods in regression.

# Appendix F: Common Misuses and Misinterpretations

The recent literature in a cross section of the life sciences suggests several problem areas. I will explain a dozen of these including my own observations along with ideas suggested by various reviewers. Related suggestions are found in Anderson and Burnham (2002). Some other comments and opinions are given at www.warnercnr.colostate.edu/~anderson/PDF_files/AIC%20Myths%20and%20Misunderstandings.pdf

1. Often too little time is devoted to generating a good set of alternative hypotheses. Some published papers seem to suggest that this important step was almost an afterthought. It might be useful for an investigator preparing to collect data to ask himself "how much effort was put into developing my specific objectives and outlining the alternative hypotheses." If the answer is "a few hours," then it might be best to revisit these important issues.

2. Some authors tend to ignore sample size issues when interpreting model selection results and then compounding this by misinterpreting the results in a dichotomous yes/no fashion (e.g., "... uptake rates did not vary across study groups" or "...there was no difference in transition probability by group"). Of course, rates of uptake and transition probabilities differed; the issue is "by how much"? Perhaps they meant to say that with the sample size available, differences in uptake seemed small. Or perhaps, the estimated differences were large, but the sample size was so small that models with such differences could not be supported.

   The lowest level of reliable inference is the sign of the effect (+ or −). If even the evidence for the sign is weak, perhaps judgment should be withheld. The parameter estimate and its confidence interval could be given but one should probably admit that the estimated effect is about 0 as far as the information in the data are concerned.

3. Some papers misinterpret the relative importance of models within about $2\Delta_i$ units when there is no change in the deviance and differing by only one parameter (the "pretending variable problem"). This issue is aggravated when the values of the maximized log-likelihoods or the deviance are not tabled (see Anderson et al. 2001b).

4. Other literature has appeared where model building, fitting, selection, and inference are treated piecemeal (e.g., splitting a dataset for purposes of "validation," treating groups, such as gender, separately without hypothesizing that some parameters might be in common across groups, ignoring the principle of parsimony in hypothesizing and modeling). These are not easy issues to understand but sometimes available software will help with this issue (e.g., the R package of freeware, Venables and Smith (2002)).

5. Some papers provide only a table of AICc and $\Delta_i$ values allowing a ranking of the models and their hypotheses. This approach might have been reasonable 10–15 years ago; however, much more can be learned using the model probabilities, evidence ratios, and model-averaged parameter estimates to gain insights into estimated effect sizes and structural relationships. In any case, inference should not stop at just identifying the "best model" as estimates of model parameters should be interpreted and these insights should be tied back to the science hypotheses.

6. A large percentage of papers present the results of simple studies as a NHT (see Stephens et al. 2005); perhaps without realizing that an evidence ratio would be easier to compute and provide a proper strength of evidence for *both* the null and the alternative (e.g., the model probabilities). The information-theoretic approach provides the probability of both the null and the

probability of the alternative, model-averaged estimates of effect size and estimates of precision that include a variance component for model selection uncertainty (see Schmidt et al. (2004) for a nice example).

7. Perhaps the worst issue is the feeling among some people that the information-theoretic methods "require" sustained thinking leading to hypothesizing good alternatives and this is too hard and too much to expect (see discussion by Steidl (2007)). Therefore, the NHT approach might be preferred because less thinking is required (i.e., one can always trump up a null). This attitude often fosters people spending resources playing the "measuring nature game" without much purpose. Alternative science hypotheses and hard thinking represent the very core of good science; good science is not always "easy." I doubt if anyone has received a Nobel Prize for testing a null hypothesis.

8. I often hear that some authors are encouraged/forced by journal editors or associate editors to add $P$-values in place of (or in addition to) estimates of effect size and their confidence intervals and model selection statistics. It seems, to me, that the peer review process could be much better. A colleague suggested that the weakest link in our science is that the accumulation of supposed knowledge is based on the unsupervised individual application of statistical hypothesis testing with very little effective oversight in the review process.

9. I see where investigators have conducted all-possible paired comparisons using $t$ tests and then used those that were statistically "significant" from the null in a multiple regression model (i.e., the "nonsignificant" variables are discarded). This is often followed by discarding the variables in the regression model that are then not "significant." This procedure attempts to "weed out" nonsignificant variables before moving to a multivariable regression analysis with a further weeding of those found to be nonsignificant once the "more comprehensive" modeling started. This strategy is not without its logic if one has no background in statistical theory and stochasticities.

   However, this strategy is very poor for several important reasons (e.g., it mixes analysis paradigms, leads to a host of technical matters such as the multiple testing problem, and often makes hidden assumptions concerning independence of the predictor variables). If the simple models represented plausible hypotheses, they should have been in the candidate set of the initial regression models. Underlying this type of error is that the focus of the investigation has improperly focused on models rather than concentrating on the science issues (i.e., plausible hypotheses). The situation points to poor study design that often stems from shallow thinking about the science issue in the first place. This problematic approach is rampant in some areas of the life sciences.

10. The use of too many models is problematic. This is often the result of a focus on running models versus thinking about the alternative science hypotheses. Certainly, if there are more models than the size of the sample

$(R > n)$, one should expect difficulties. Large, unfocused descriptive studies are often faced with a huge number of models (see no. 11, below).
11. Some software packages allow one to perform a "stepwise AIC" and this represents poor practice. The theme here seems to be that the computer will "find out what is important without the investigator having to think." The underlying problem, like running "all possible models," is the finding of effects that are, in fact, spurious. This issue relates back to Freedman's paradox and model selection bias. Admittedly, these are issues that are not easy to understand without some background.
12. In general, I think the results from rampant data dredging should often remain unpublished. I think more should be done to explain to readers what results and conclusions stem from a priori considerations versus the more tentative insights from *post hoc* investigations. Such statements portray honesty and openness in publication and can help define the next set of hypotheses and their models.

# References

Abelson, R. P. (1995). *Statistics as principled argument*. Lawrence Erlbaum Associates, Hillsdale, NJ.

Akaike, H. (1973). Information theory as an extension of the maximum likelihood principle. *in* B. N. Petrov, and F. Csaki (Eds.) *Second International Symposium on Information Theory*. Akademiai Kiado, Budapest. pp. 267–281.

Akaike, H. (1974). A new look at the statistical model identification. *IEEE Transactions on Automatic Control AC* **19**, 716–723.

Akaike, H. (1977). On entropy maximization principle. *in* P. R. Krishnaiah (Ed.) *Applications of statistics*. North-Holland, Amsterdam. pp. 27–41.

Akaike, H. (1978). A Bayesian analysis of the minimum AIC procedure. *Annals of the Institute of Statistical Mathematics* **30**, 9–14.

Akaike, H. (1981a). Likelihood of a model and information criteria. *Journal of Econometrics* **16**, 3–14.

Akaike, H. (1981b). Modern development of statistical methods. *in* P. Eykhoff (Ed.) *Trends and progress in system identification*. Pergamon Press, Paris. pp. 169–184.

Akaike, H. (1983a). Statistical inference and measurement of entropy. *in* G. E. P. Box, T. Leonard, and C-F. Wu (Eds.) *Scientific inference, data analysis, and robustness*. Academic Press, London. pp. 165–189.

Akaike, H. (1983b). Information measures and model selection. *International Statistical Institute* **44**, 277–291.

Akaike, H. (1985). Prediction and entropy. *in* A. C. Atkinson, and S. E. Fienberg (Eds.) *A celebration of statistics*. Springer, New York, NY. pp. 1–24.

Akaike, H. (1987). Factor analysis and AIC. *Psychometrika* **52**, 317–332.

Akaike, H. (1992). Information theory and an extension of the maximum likelihood principle. *in* S. Kotz, and N. L. Johnson (Eds.) *Breakthroughs in statistics, Vol. 1*. Springer-Verlag, London. pp. 610–624.

Akaike, H. (1994). Implications of the informational point of view on the development of statistical science. *in* H. Bozdogan, (Ed.) *Engineering and Scientific Applications, Vol. 3*, Proceedings of the First US/Japan Conference on the Frontiers of Statistical Modeling: An Informational Approach. Kluwer, Dordrecht, Netherlands. pp. 27–38.

Anderson, D. R. (2001). The need to get the basics right in wildlife field studies. *Wildlife Society Bulletin* **29**, 1294–1297.

Anderson, D. R., and Burnham, K. P. (1999). General strategies for the collection and analysis of ringing data. *Bird Study* **46** (Supplement), S261–S270.

Anderson, D. R., and Burnham, K. P. (2002). Avoiding pitfalls when using information-theoretic methods. *Journal of Wildlife Management* **66**, 912–918.

Anderson, D. R., Burnham, K. P., Franklin, A. B., Gutierrez, R. J., Forsman, E. D., Anthony, R. G., White, G. C., and Shenk, T. M. (1999). A protocol for conflict resolution in analyzing empirical data related to natural resources controversies. *Wildlife Society Bulletin* **27**, 1050–1058.

Anderson, D. R., Burnham, K. P., Gould, W. R., and Cherry, S. (2001a). Concerns about finding effects that are actually spurious. *Wildlife Society Bulletin* **29**, 311–316.

Anderson, D. R., Burnham, K. P., and Thompson, W. L. (2000). Null hypothesis testing: Problems, prevalence, and an alternative. *Journal of Wildlife Management* **64**, 912–923.

Anderson, D. R., Burnham, K. P., and White, G. C. (1994). AIC model selection in overdispersed capture–recapture data. *Ecology* **75**, 1780–1793.

Anderson, D. R., Burnham, K. P., and White, G. C. (1998). Comparison of AIC and CAIC for model selection and statistical inference from capture–recapture studies. *Journal of Applied Statistics* **25**, 263–282.

Anderson, D. R., Burnham, K. P., and White, G. C. (2001). Kullback-Leibler information in resolving natural resource conflicts when definitive data exist. *Wildlife Society Bulletin* **29**, 1260–1270.

Anderson, D. R., Link, W. A., Johnson, D. K., and Burnham, K. P. (2001b). Suggestions for presenting the results of data analysis. *Journal of Wildlife Management* **65**, 373–378.

Anderson, S., Auquier, A., Hauck, W. W., Oakes, D., Vandaele, W., and Weisberg, H. I. (1980). *Statistical methods for comparative studies*. John Wiley, New York, NY.

Anthony, R. G., Forsman, E. D., Franklin, A. B., Anderson, D. R., Burnham, K. P., White, G. C., Schwarz, C. J., Nichols, J. D., Hines, J. E., Olson, G. S., Ackers, S. H., Andrews, L. S., Biswell, B. L., Carlson, P. C., Diller, L. V., Dugger, K. M., Fehring, K. E., Fleming, T. L., Gerhardt, R. P., Gremel, S. A., Gutierrez, R. J., Harpe, P. J., Herter, D. R., and Higley, J. M. (2006). Status and trends in demography of Northern Spotted Owls, 1985–2003. *Wildlife Monograph* **163**, 1–48.

Atilgan, T. (1996). Selection of dimension and basis for density estimation and selection of dimension, basis and error distribution for regression. *Communications in Statistics – Theory and Methods* **25**, 1–28.

Atmar, W. 2001. A profoundly repeated pattern. *Bulletin of the Ecological Society of America* **26**, 208–211.

Azzalini, A. (1996). *Statistical inference based on the likelihood*. Chapman and Hall, London, UK.

Ball, L. C., Doherty, P. F., Jr., and McDonald, M. W. (2005). An occupancy modeling approach to evaluating Palm Springs ground squirrel habitat model. *Journal of Wildlife Management* **69**, 894–904.

Bedrick, E. J., and Tsai, C.-L. (1994). Model selection for multivariate regression in small samples. *Biometrics* **50**, 226–231.

Blanckenhorn, W. U., Hellriegel, B., Hosken, D. J., Jann, P., Altweg, R., and Ward, P. I. (2004). Does testis size track expected mating success in yellow dung flies? *Functional Ecology* **18**, 414–418.

Boltzmann, L. (1877). Uber die Beziehung zwischen dem Hauptsatze der mechanischen Warmetheorie und der Wahrscheinlicjkeitsrechnung respective den Satzen uber das Warmegleichgewicht. *Wiener Berichte* **76**, 373–435.

Bortz, D. M., and Nelson, P. W. (2006). Model selection and mixed-effects modeling of HIV infection dynamics. *Bulletin of Mathematical Biology* **68**, 2005–2025.

Box, G. E. P. (1976). Science and statistics. *Journal of the American Statistical Association* **71**, 791–799.

Box, G. E. P. (1979). Robustness in scientific model building. *in* R. L. Launer and G. N. Wilkinson, eds., *Robustness in statistics*. Academic Press, New York, NY. pp. 201–236.

Box, G. E. P., Leonard, T., and Wu, C.-F. (Eds.) (1981). *Scientific inference, data analysis, and robustness*. Academic Press, London.

Box, J. F. (1978). *R. A. Fisher: the life of a scientist*. John Wiley, New York, NY. pp. 511.

Bozdogan, H. (1987). Model selection and Akaike's information criterion (AIC): The general theory and its analytical extensions. *Psychometrika* **52**, 345–370.

Breiman, L. (1992). The little bootstrap and other methods for dimensionality selection in regression: $X$-fixed prediction error. *Journal of the American Statistical Association* **87**, 738–754.

Breiman, L., and Freedman, D. F. (1983). How many variables should be entered in a regression equation? *Journal of the American Statistical Association* **78**, 131–136.

Broda, E. (1983). *Ludwig Boltzmann: Man, physicist, philosopher.* (translated with L. Gay). Ox Bow Press, Woodbridge, Connecticut, USA.

Brown, D., and Rothery, P. (1993). *Models in biology: Mathematics, statistics and computing.* John Wiley, New York, NY.

Brownie, C., Anderson, D. R., Burnham, K. P., and Robson, D. S. (1985). *Statistical inference from band recovery data – a handbook.* 2nd Ed. U. S. Fish and Wildlife Service Resource Publication 156. pp. 305.

Brush, S. G. (1965). *Kinetic theory, Vol. 1.* Pergamon Press, Oxford.

Brush, S. G. (1966). *Kinetic theory, Vol. 2.* Pergamon Press, Oxford.

Buckland, S. T., Burnham, K. P., and Augustin, N. H. (1997). Model selection: An integral part of inference. *Biometrics* **53**, 603–618.

Burnham, K. P., and Anderson, D. R. (1992). Data-based selection of an appropriate biological model: the key to modern data analysis. *in* D. R. McCullough, and R. H. Barrett (Eds.) *Wildlife 2001: Populations.* Elsevier, London. pp. 16–30.

Burnham, K. P., and Anderson, D. R. (2001). Kullback-Leibler information as a basis for strong inference in ecological studies. *Wildlife Research* **28**, 111–119.

Burnham, K. P., and Anderson, D. R. (2002). *Model selection and multimodel inference: A practical information-theoretic approach, 2nd Ed.*, Springer-Verlag, New York, NY.

Burnham, K. P., and D. R. Anderson. (2004). Multimodel inference: Understanding AIC and BIC in model selection. *Sociological Methods and Research* **33**, 261–304.

Burnham, K. P., Anderson, D. R., and White, G. C. (1994). Evaluation of the Kullback–Leibler discrepancy for model selection in open population capture-recapture models. *Biometrical Journal* **36**, 299–315.

Burnham, K. P., Anderson, D. R., and White, G. C. (1995b). Selection among open population capture–recapture models when capture probabilities are heterogeneous. *Journal of Applied Statistics* **22**, 611–624.

Burnham, K. P., Anderson, D. R., and White, G. C. (1996). Meta-analysis of vital rates of the Northern Spotted Owl. *Studies in Avian Biology* **17**, 92–101.

Burnham, K. P., Anderson, D. R., White, G. C., Brownie, C., and Pollock, K. H. (1987). *Design and analysis methods for fish survival experiments based on release-recapture.* American Fisheries Society, Monograph **5**, 437.

Burnham, K. P., White, G. C., and Anderson, D. R. (1995a). Model selection in the analysis of capture-recapture data. *Biometrics* **51**, 888–898.

Caley, P., and Hone, J. (2002). Estimating the force of infection; *Mycobacterium bovis* infection in feral ferrets *Mustela furo* in New Zealand. *Journal of Animal Ecology* **71**, 44–54.

Caley, P., and Hone, J. (2005). Assessing the host disease status of wildlife and the implications for disease control: *Mycobacterium bovis* infection in feral ferrets. *Journal of Animal Ecology* **42**, 708–719.

Carlin, B. P., and Louis, T. A. (2001). *Bayes and empirical Bayes methods for data analysis, 2nd Ed.* Chapman and Hall, New York, NY.

Carter, G. M., Stolen, E. D., and Breininger, D. R. (2006). A rapid approach to modeling species–habitat relationships. *Biological Conservation* **127**, 237–244.

Chamberlain, T. C. (1890). The method of multiple working hypotheses. *Science* **15**, 92–96. (Reprinted 1965, *Science* **148**, 754–759.)

Chatfield, C. (1991). Avoiding statistical pitfalls (with discussion). *Statistical Science* **6**, 240–268.

Chatfield, C. (1995a). *Problem solving: A statistician's guide.* Chapman and Hall, London. pp. 325.

Chatfield, C. (1995b). Model uncertainty, data mining and statistical inference. *Journal of the Royal Statistical Society, Series A* **158**, 419–466.

Chen, M.-H., Shao, Q.-M., and Ibrahim, J. G. (2000). *Monte Carlo methods in Bayesian computation.* Springer, New York, NY.

Clyde, M. (2000). Model uncertainty and health effect studies for particular matter. *Environmetrics* **11**, 745–763.

Cohen, D. (1966). Optimizing reproduction in a randomly varying environment. *Journal of Theoretical Biology* **12**, 119–129.

Cohen, D. (1967). Optimizing reproduction in a randomly varying environment when a correlation may exist between the conditions at the time a choice has to be made and the subsequent outcome. *Journal of Theoretical Biology* **16**, 1–14.

Cohen, D. (1968). A general model of optimal reproduction in a randomly varying environment. *Journal of Ecology* **56**, 219–228.

Cohen, E. G. D., and Thirring, W. (Eds.) (1973). *The Boltzmann equation: Theory and applications*. Springer-Verlag, New York, NY. pp. 642.

Cohen, J., and Medley, G. (2005). *Stop working & start thinking*. Taylor & Francis Group, New York, NY.

Cook. T. D, and Campbell, D. T. (1979). *Quasi-experimentation: Design and analysis issues for field settings*. Houghton Mifflin Company, Boston, MA.

Cover, T. M., and Thomas, J. A. (1991). *Elements of information theory*. John Wiley, New York, NY. pp. 542.

Cox, D. R. (1990). Role of models in statistical analysis. *Statistical Science* **5**, 169–174.

Cox, D. R. (1995). The relation between theory and application in statistics. *Test* **4**, 207–261.

Cox, D. R. (2006). *Principles of statistical inference*. Cambridge University Press, Cambridge, UK.

Daniel, C., and Wood, F. S. (1971). *Fitting equations to data*. Wiley-Interscience, New York, NY. pp. 342.

Dawkins, R. (1986). *The blind watchmaker: Why the evidence of evolution reveals a universe without design*. W. W. Norton, New York, NY.

deLeeuw, J. (1992). Introduction to Akaike (1973) information theory and an extension of the maximum likelihood principle. *in* S. Kotz, and N. L. Johnson (Eds.) *Breakthroughs in statistics, Vol. 1*. Springer-Verlag, London. pp. 599–609.

Delury, D. B. (1954). On the assumptions underlying estimates of mobile populations. Pages 287–293 *in Statistics and Mathematics in biology*. Iowa State University, Ames.

Dijkstra, T. K. (Ed). (1988). *On model uncertainty and its statistical implications*. Lecture Notes in Economics and Mathematical Systems, Springer-Verlag, New York, NY. pp. 138.

Draper, D. (1995). Assessment and propagation of model uncertainty (with discussion). *Journal of the Royal Statistical Society, Series B* **57**, 45–97.

Draper, N. R., and Smith, H. (1981). *Applied regression analysis*. John Wiley, New York, NY. pp. 709.

Eberhardt, L. L., and Thomas, J. M. (1991). Designing environmental field studies. *Ecological Monographs* **61**, 53–73.

Edwards, A. W. F. (1972). *Likelihood*. Cambridge University Press, Cambridge, UK.

Edwards, A. W. F. (1976). *Likelihood: An account of the statistical concept of likelihood and its application to scientific inference*. Cambridge University Press, Cambridge. pp. 235.

Edwards, A. W. F. (1992). *Likelihood: Expanded edition*. Johns Hopkins University Press, Baltimore, Maryland.

Edwards, A. W. F. (2001). Occam's bonus. p. 128–139; in Zellner, A., Keuzenkamp, H. A., and McAleer, M. *Simplicity, inference and modelling*. Cambridge University Press, Cambridge, UK.

Elliott, L. P., and Brook, B. W. (2007). Revisiting Chamberlin (1890): Multiple working hypotheses for the 21st century. *Bioscience*, **57**, 608–614.

Eng, J. (2004). Sample size estimation: A glimpse beyond simple formulas. *Radiology* **230**, 606–612.

Everitt, B. S. (1998). *The Cambridge dictionary of statistics*. Cambridge University Press, Cambridge, UK.

Findley, D. F., and Parzen, E. (1995). A conversation with Hirotugu Akaike. *Statistical Science* **10**, 104–117.

Fisher, R. A. (1936). Uncertain inference. *Proceedings of the American Academy of Arts and Sciences* **71**, 245–258.

Flack, V. F., and Chang, P. C. (1987). Frequency of selecting noise variables in subset regression analysis: A simulation study. *The American Statistician* **41**, 84–86.

Flather, C. H. (1996). Fitting species-accumulation functions and assessing regional land use impacts on avian diversity. *Journal of Biogeograhy* **23**, 155–168.

Ford, E. D. (2000). *Scientific method for ecological research*. Cambridge University Press, Cambridge, UK.

Forsche, B. K. (1963). Chaos in the brickyard. Science **142**, 339.

Freddy, D. J., White, G. C., Kneeland, M. C., Kahn, R. H., Unsworth, J. W., deVergie, W. J., Graham, V. K., Ellenberger, J. H., and Wagner, C. H. (2004). How many mule deer are there? Challenges of credibility in Colorado. *Wildlife Society Bulletin* **32**, 916–927.

Freedman, D. A. (1983). A note on screening regression equations. *The American Statistician* **37**, 152–155.

Freedman, D. A., Navidi, W., and Peters, S. C. (1988). On the impact of variable selection in fitting regression equations. *in* T. K. Dijkstra (Ed.) *On model uncertainty and its statistical implications*. Lecture Notes in Economics and Mathematical Systems, Springer-Verlag, New York, NY. pp. 1–16.

Fujikoshi, Y., and Satoh, K. (1997). Modified AIC and $C_p$ in multivariate linear regression. *Biometrika* **84**, 707–716.

Gallager, R. G. (2001). Claude E. Shannon: A retrospective on his life, work, and impact. *IEEE Transactions on Information Theory* **47**, 2681–2695.

Gelman, A., Carlin, J. B., Stern, H. S., and Rubin, D. B. (2003). *Bayesian data analysis, 2nd Ed.* Chapman and Hall, New York, NY.

Gilchrist, W. (1984). *Statistical modelling*. Chichester, Wiley and Sons, New York, NY.

Givens, G. H., and Hoeting, J. A. (2005). *Computational statistics*. John Wiley, Hoboken, NJ.

Goldman, S. (1953). *Information theory*. Constable Publishing, London, UK.

Golomb, S. W., Berlekamp, E., Cover, T. M. Gallager, R. G., Massey, J. L., and Viterbi, A. J. (2002). Claude Elwood Shannon. *Notices of American Mathematical Society* **292**, 8–16.

Golub, G. H., Health, M., and Wahba, G. (1979). Generalized cross validation as a method for choosing a good ridge parameter. *Technometrics* **21**, 215–223.

Good, I. J. (1979). A. M. Turing's statistical work in World War II. *Biometrika* **66**, 393–396.

Goodman, S. N., and Royall, R. (1988). Evidence and scientific research. *American Journal of Public Health* **78**, 1568–1574.

Gotelli, N. J., and Ellison, A. M. (2004). *A primer of ecological statistics*. Sinauer Associates, Sunderland, MA.

Greenhouse, S. W. (1994). Solomon Kullback: 1907–1994. *Institute of Mathematical Statistics Bulletin* **23**, 640–642.

Guiasu, S. (1977). *Information theory with applications*. McGraw-Hill, New York, NY.

Gurka, M. J. (2006). Selecting the best linear mixed model under REML. *American Statistician* **60**, 19–26.

Hairston, N. G. (1989). *Ecological experiments: Purpose, design and execution*. Cambridge University Press, Cambridge, UK.

Hald, A. (1952). *Statistical theory with engineering applications*. John Wiley, New York, NY.

Hand, D. J. (1994). Statistical strategy: Step 1. *in* P. Cheeseman, and R. W. Oldford (Eds.) *Selecting models from data*. Springer-Verlag, New York, NY. pp. 1–9.

Hasenöhrl, F. (Ed.) (1909). *Wissenschaftiche Abhandlungen*. 3 Vols, Leipzig, Germany.

Hendry, A. P., Grant, P. R., Grant, B. R., Ford, H.A., Brewer, M. J., and Podos, J. (2006). Possible human impacts on adaptive radiation: Beak size bimodality in Darwin's finches. *Proceedings of the Royal Society, Series B*. Published online. 1–8.

Hilborn, R., and Mangle, M. (1997). *The ecological detective*. Princeton University Press, Princeton, NJ.

Hjorth, J. S. U. (1994). *Computer intensive statistical methods: Validation, model selection and bootstrap*. Chapman and Hall, London.

Hobbs, N. T., and Hilborn, R. (2006). Alternatives to statistical hypothesis testing in ecology: A guide to self teaching. *Ecological Applications* **16**, 5–19.

Hobson, A., and Cheng, B.-K. (1973). A comparison of the Shannon and Kullback information measures. *Journal of Statistical Physics* **7**, 301–310.

Hoeting, J. A., Madigan, D., Reftery, A. E., and Volinsky, C. T. (1999). Bayesian model averaging: A tutorial (with discussion). *Statistical Science* **14**, 382–417.

Hoeting, J. A., Davis, R. A., Merton, A. A., and Thompson, S. E. (2006). Model selection for geostatistical models. *Ecological Applications* **16**, 87–98.

Horner, C., and Westacott, E. (2000). *Thinking through philosophy: An introduction*. Cambridge University Press, Cambridge, UK.

Hurvich, C. M., Simonoff, J. S., and Tsai, C.-L. (1998). Smoothing parameter selection in nonparametric regression using an improved Akaike information criterion. *Journal of the Royal Statistical Society Series B*, **60**, 271–293.

Hurvich, C. M., and Tsai, C.-L. (1989). Regression and time series model selection in small samples. *Biometrika* **76**, 297–307.

Hurvich, C. M., and Tsai, C-L. (1990). The impact of model selection on inference in linear regression. *The American Statistician* **44**, 214–217.

Hurvich, C. M., and Tsai, C-L. (1991). Bias of the corrected AIC criterion for underfitted regression and time series models. *Biometrika* **78**, 499–509.

Hurvich, C. M., and Tsai, C-L. (1995). Model selection for extended quasi-likelihood models in small samples. *Biometrics* **51**, 1077–1084.

Ishiguro, M., Sakamoto, Y., and Kitagawa, G. (1997). Bootstrapping log likelihood and EIC, an extension of AIC. *Annals of the Institute of Statistical Mathematics* **29**, 411–434.

Jaynes, E. T. (1957). Information theory and statistical mechanics. *Physics Review* **106**, 620–630.

Jessop, A. (1995). *Informed assessments: An introduction to information, entropy and statistics.* Ellis Horwood, London. pp. 366.

Johnson, D. L. (1999). The insignificance of statistical significance. *Journal of Wildlife Management*, **63**, 763–772.

Johnson, J. W. (1996). Fitting percentage of body fat to simple body measurements. *Journal of Statistics Education* **4** ( e-journal).

Karban, R., and Huntzinger, M. (2006). How to do ecology: a concise handbook. Princeton University Press, Princeton, NJ.

Kendall, W. L., and Gould, W. R. (2002). An appeal to undergraduate wildlife programs: Send scientists to learn statistics. *Wildlife Society Bulletin* **30**, 623–627.

Keppie, D. M. (2006). Context, emergence, and research design. *Wildlife Society Bulletin* **34**, 242–246.

Kitagawa, T. (1986). Editor's preface. *in* Y. Sakamoto, Ishiguro, M., and Kitagawa, G. *Akaike information criterion statistics.* KTK Scientific Publishing Company, Tokyo, Japan. pp. xiii–xiv.

Konishi, S., and Kitagawa, G. (2007). *Information criteria and statistical modeling.* Springer, New York, NY.

Krebs, C. J. (2000). Hypothesis testing in ecology. *in* L. Boitani and T. K. Fuller, *Research techniques in animal ecology: Controversies and consequences.* Columbia University Press, New York, NY. pp. 1–12.

Kullback, S. (1959). *Information theory and statistics.* John Wiley, New York, NY.

Kullback, S., and Leibler, R. A. (1951). On information and sufficiency. *Annals of Mathematical Statistics* **22**, 79–86.

Kullback, S. (1987). The Kullback-Leibler distance. *The American Statistician* **41**, 340–341.

Kuhn, T. S. (1970). *The structure of scientific revolutions, 2nd Ed.* University of Chicago Press, Chicago, IL.

Kutner, M. H., Nachtsheim, C. J., Neter, J., and Li, W. (2004). *Applied linear statistical models, 4th Ed.* McGraw Hill, Chicago, IL.

Lahiri, P. (Ed.) (2001). *Model selection.* Institute of Mathematical Statistics, Lecture Note – Monograph Series **38**, 256.

Leamer, E. E. (1978). *Specification searches: Ad hoc inference with nonexperimental data.* John Wiley, New York, NY.

Lebreton, J.-D., Burnham, K. P., Clobert, J., and Anderson, D. R. (1992). Modeling survival and testing biological hypotheses using marked animals: A unified approach with case studies. *Ecological Monograph* **62**, 67–118.

Lee, Y., Nelder, J. A., and Pawitan, Y. (2006). *Generalized linear models with random effects: Unified analysis via H-likelihood.* Chapman and Hall, Boca Raton, FL.

Lehmann, E. L. (1990). Model specification: The views of Fisher and Neyman, and later developments. *Statistical Science* **5**, 160–168.

Leopold, A. (1933) *Game management.* University of Wisconsin Press, Madison, WI.

Levins, R. (1966). The strategy of model building in population biology. *American Scientist* **54**, 421–431.

Linhart, H., and Zucchini, W. (1986). *Model selection*. John Wiley, New York, NY.

Link, W. A., and Barker, R. J. (2006). Models weights and the foundations of multimodel inference. *Ecology* **87**, 2626–2635.

Lukacs, P. M., Thompson, W. L., Kendall, W. L., Gould, W. R., Doherty, P. F., Burnham, K. P., and Anderson, D. R. (2007). Concerns regarding a call for pluralism of information theory and hypothesis testing. *Journal of Animal Ecology* **44**, 456–460.

Lukacs, P. M., Burnham, K. P., and Anderson, D. R. (Ed.). Freedman's paradox: Why ecologists should worry (unpublished).

Lunneborg, C. E. (1994). *Modeling experimental and observational data*. Duxbury Press, Belmont, CA, USA. pp. 506.

Mallows, C. L. (1973). Some comments on $C_p$. *Technometrics*, **12**, 591–612.

MacKenzie, D. I., Nichols, J. D., Royle, J. A., Pollock, K. H., Bailey, L. L., and Hines, J. E. (2006). *Occupancy estimation and modeling: Inferring patterns and dynamics of species occurrence*. Elsevier, London, UK.

Manly, B. F. J. (1992). *The design and analysis of research studies*. Cambridge University Press, Cambridge, UK.

Massart, P. (2007). *Concentration inequalities and model selection*. Springer-Verlag, Berlin.

Mauer, B. A. (1999). *Untangling ecological complexity*. University of Chicago Press, Chicago, IL.

McCullagh, P., and Nelder, J. A. (1989). *Generalized linear models, 2nd Ed*. Chapman and Hall, New York, NY.

McCulloch, C. E. (2003). *Generalized linear models*. NSF-CBMS Regional Conference Series in Probability and Statistics **7**, 1–84.

McQuarrie, A. D. R., and Tsai, C.-L. (1998). *Regression and time series model selection*. World Scientific, London, UK.

Mead, R. (1988). *The design of experiments: Statistical principles for practical applications*. Cambridge University Press, New York, NY.

Miller, A. J. (2002). *Subset selection in regression, 2nd Ed*., Chapman and Hall, London, UK.

Moore, B. N., and Parker, R. (1986). *Critical thinking, 5th Ed*. Mayfield Publishing Company, London, UK.

Morgan, B. J. T. (2000). *Applied stochastic modelling*. Arnold Publishing, London, UK.

Muthen, L. K., and Muthen, B. O. (2002). How to use a Monte Carlo study to decide on sample size and determine power. *Structural Equation Modeling* **9**, 599–620.

Nagelkerke, N. J. K. (1991). A note on a general definition of the coefficient of determination. *Biometrika* **78**, 691–692.

Nelder, J. A. (1991). Generalized linear models for enzyme-kinetic data. *Biometrics* **47**, 1605–1615.

Nichols, J. D. (2001). Using models in the conduct of science and management of natural resources. *in* Shenk, T., and Franklin, A. B. (Eds.) *Modeling in natural resource management: Development, interpretation, and application*. Island Press, Washington, D. C. pp. 11–34.

O'Connor, R. J. (2000). Why ecology lags behind biology. *The Scientist* **14**, 35.

Oliver, J. E. (1991). *The incomplete guide to the art of discovery*. Columbia University Press, New York, NY.

Pan, W. (2001a). Akaike's information criterion in generalized estimating equations. *Biometrics* **57**, 120–125.

Pan, W. (2001b). Model selection in estimating equations. *Biometrics* **57**, 529–534.

Parzen, E. (1994). Hirotugu Akaike, statistical scientist. *in* H. Bozdogan (Ed.) *Engineering and Scientific Applications, Vol. 1*, Proceedings of the First US/Japan Conference on the Frontiers of Statistical Modeling: An Informational Approach. Kluwer Academic Publishers, Dordrecht, Netherlands. pp. 25–32.

Parzen, E., Tanabe, K., and Kitagawa, G. (Eds.) (1998). *Selected papers of Hirotugu Akaike*. Springer-Verlag, New York, NY.

Pawitan, Y. (2001). *In all likelihood: Statistical modelling and inference using likelihood*. Oxford Science Publications, Oxford, UK.

Peirce, C. S. (1955). Abduction and induction. *in* J. Buchler (Ed.) *Philosophical writings of Peirce*. Dover, New York, NY. pp. 150–156.

Pigliucci, M. (2002a). Are ecology and evolutionary biology "soft" sciences? *Annals of Zoological Fennici* **39**, 87–98.

Pigliucci, M. (2002b). *Denying evolution: Creationism, scientism, and the nature of science*. Sinauer Associates, Sunderland, MA.

Pistorius, P. A., Bester, M. N., Kirkman, S. P., and Boveng, P. L. (2000). Evaluation of age- and sex-dependent rates of tag loss in southern elephant seals. *Journal of Wildlife Management* **64**, 373–380.

Platt, J. R. (1964). Strong inference. *Science* **146**, 347–353.

Popper, K. R. (1959). *The logic of scientific discovery*. Harper and Row, New York, NY.

Potscher, B. M. (1991). Effects of model selection on inference. *Econometric Theory* **7**, 163–185.

Qin, J., and Lawless, G. (1994). Empirical likelihood and general estimating equations. *Annals of Statistics* **22**, 300–325.

Rao, C. R. (2004). Forward. *in* Taper, M. L., and Lele, S. R. *The nature of scientific evidence: Statistical, philosophical, and empirical considerations*. University of Chicago Press, Chicago, IL. pp. xi–xiii.

Remontet, L., Bossard, N., Belot, A., Esteve, J., and the French network of cancer registries. (2006). An overall strategy based on regression models to estimate relative survival and model the effects of prognostic factors in cancer survival studies. *Statistics in Medicine* **26**, 2214–2228.

Rencher, A. C., and Pun, F. C. (1980). Inflation of $R^2$ in best subset regression. *Technometrics* **22**, 49–53.

Renshaw, E. (1991). *Modelling biological populations in space and time*. Cambridge University Press, Cambridge, UK.

Reschenhofer, E. (1996). Prediction with vague prior knowledge. *Communications in Statistics – Theory and Methods* **25**, 601–608.

Resetarites, W. J., Jr., and Bernardo, J. (Eds.) (2001). *Experimental ecology: Issues and perspectives*. Oxford University Press, UK.

Rissanen, J. (1989). *Stochastic complexity in statistical inquiry*. World Scientific, Series in Computer Science, Vol 15, Singapore.

Rissanen, J. (1996). Fisher information and stochastic complexity. *IEEE Transactions on Information Theory* **42**, 40–47.

Rissanen, J. (2007). Information and complexity in statistical modeling. Springer, New York, NY.

Rosenbaum, P. R. (2002). *Observational studies, 2nd Ed.*, Springer-Verlag, New York, NY.

Royall, R. M. (1997). *Statistical evidence: A likelihood paradigm*. Chapman and Hall, London, UK.

Romesburg, H. C. (2002). *The life of the creative spirit*. Xlibris Corporation, Rome, IT.

Sakamoto, Y. (1991). *Categorical data analysis by AIC*. KTK Scientific Publishers, Tokyo.

SAS Institute Inc. (2004). *SAS® Language guide for personal computers*, Version 9.1 Edition. SAS Institute Inc, Cary, North Carolina.

SAS Institute. (2004). SAS/STAT® user's guide. 6th Ed. SAS Institute, Cary, NC.

Scheiner, S. M., and Gurevitch, J. (Eds.) (1993). *Design and analysis of ecological experiments*. Chapman and Hall, London.

Schmidt, B. R., Feldmann, R., and Schaub, M. (2004). Demographic processes underlying population growth and decline in *Salamandra salamandra*. *Conservation Biology* **19**, 1149–1156.

Schwarz, G. (1978). Estimating the dimension of a model. *Annals of Statistics* **6**, 461–464.

Sclove, S. L. (1987). Application of some model-selection criteria to some problems in multivariate analysis. *Psychometrika* **52**, 333–343.

Seghouane, A.-K. (2005). Multivariate model selection with KIC for extrapolated cases. *Neural Networks*, Proceedings, 2005 IEEE International Joint Conference **2**, 1292–1295.

Seghouane, A.-K. (2006). Multivariate regression model selection from small samples using Kullback's symmetric divergence. *Signal Processing* **86**, 2074–2084.

Severini, T. A. (2000). *Likelihood methods in statistics*. Oxford University Press, Oxford, UK.

Shadish, W. R., Cook, T., and Campbell, D. T. (2002). *Experimental and quasi-experimental designs for generalized causal inference*. Houghton Mifflin Company, New York, NY.

Shannon, C. E. (1948). A mathematical theory of communication. *Bell System Technical Journal* **27**, 379–423 and 623–656.

Shenk, T., and Franklin, A. B. (2001). *Modeling in natural resource management: Development, interpretation, and application*. Island Press, Washington, D. C.

Shi, R., and Tsai, C.-L. (2002). Regression model selection – a residual likelihood approach. *Journal of the Royal Statistical Association, Series B*, **64**, 237–252.

Siotani, M., and Wakaki, H. (2006). Contribution to multivariate analysis by Professor Yasunori Fujikoshi. *Journal of Multivariate Analysis* **97**, 1914–1926.

Smith, D. L., Dushoff, J., Snow, R. W., and Hay, S. I. (2005). The entomological inoculation rate and *Lasmodium falciparum* infection in African children. *Nature, letters*, 04024.

Snedecor, G. W., and Cochran, W. G. (1989). *Statistical methods, 8th Ed*. Iowa State University Press, Ames.

Soofi, E. S. (1994). Capturing the intangible concept of information. *Journal of the American Statistical Association* **89**, 1243–1254.

Soule, M. E. (1987). Where do we go from here? *in* M. E. Soule (Ed.) *Viable populations for conservation.* Cambridge University Press, Cambridge, UK. pp. 175–183

Speed, T. P., and Yu, B. (1993). Model selection and prediction: Normal regression. *Annals of the Institute of Statistical Mathematics* **1**, 35–54.

Spiegelhalter, D. J., Best, N. G., Carlin, B. P., and van der Linde, A. (2002). Bayesian measures of model complexity and fit. *Journal of the Royal Statistical Society* **64** (3), 1–34.

Starfield, A. M., and Bleloch, A. L. (1991). *Building models for conservation and wildlife management, 2nd Ed.* Burgess Press, Edina, MN.

Starfield, A. M., Smith, K. A., and Bleloch, A. L. (1990). *How to model it: Problem-solving for the computer age.* McGraw-Hill, New York, NY.

Steidl, R. J. (2007). Model selection, hypothesis testing, and risks of condemning analytical tools. *Journal of Wildlife Management* **70**, 1497–1498.

Stephens, P. A., Buskirk, S. W., Hayward, G. D., and Martinez del Rio, C. (2005). Information theory and hypothesis testing: A call for pluralism. *Journal of Applied Ecology* **42**, 4–12.

Stone, M. (1974). Cross-validatory choice and assessment of statistical predictions (with discussion). *Journal of the Royal Statistical Society, Series B* **39**, 111–147.

Stone, M. (1977). An asymptotic equivalence of choice of model by cross-validation and Akaike's criterion. *Journal of the Royal Statistical Society, Series B* **39**, 44–47.

Sugiura, N. (1978). Further analysis of the data by Akaike's information criterion and the finite corrections. *Communications in Statistics, Theory and Methods.* **A7**, 13–26.

Swihart, R. K., Dunning, J. B., Jr., and Waser, P. M. (2002). Gray matters in ecology: Dynamics of pattern, process, and scientific progress. *Bulletin of the Ecological Society of America* **83**, 149–155.

Takeuchi, K. (1976). Distribution of informational statistics and a criterion of model fitting. *Suri-Kagaku* (Mathematic Sciences) **153**, 12–18 (In Japanese).

Taper, M. L., and Lele, S. R. (2004). *The nature of scientific evidence: statistical, philosophical, and empirical considerations.* University of Chicago Press, Chicago, IL.

Taubes, G. (1995). Epidemiology faces its limits. *Science* **269**, 164–169.

Thomas, L., Laake, J. L., Strindberg, S., Margues, F. F. C., Buckland, S. T., Borchers, D. L., Anderson, D. R., Burnham, K. P., Hedley, S. L., Pollard, J. H., Bishop, J. R. B., and Marques, T. A. (2006). Distance 5.0. Release 2. Research Unit for Wildlife Population Assessment, University of St. Andrews, UK.

Tong, H. (1994). Akaike's approach can yield consistent order determination. Pages 93–103 *in* H. Bozdogan (Ed.) *Engineering and Scientific Applications.* Vol. 1, Proceedings of the First US/Japan Conference on the Frontiers of Statistical Modeling: An Informational Approach. Kluwer Academic Publishers, Dordrecht, Netherlands.

Ullah, A. (1996). Entropy, divergence and distance measures with econometric applications. *Journal of Statistical Planning and Inference* **49**, 137–162.

Umbach, D. M., and Wilcox, A. J. (1996). A technique for measuring epidemiologically useful features of birthweight distributions. *Statistics in Medicine* **15**, 1333–1348.

Van Buskirk, J., and Arioli, M. (2002). Dosage response of an induced defense: How sensitive are tadpoles to predation risk? *Ecology* **83**, 1580–1585.

van der Linde, A. (2004). On the association between a random parameter and an observable. *Test* **13**, 85–111.

Venables, W. N., and Smith, D. M. (2002). *An introduction to R*. Network Theory Limited Publishing, Bristol, UK.

Vieland, V. J., and Hodge, S. E. (1998). Review of *Statistical Evidence: A Likelihood Paradigm*. By R. Royall. *American Journal of Human Genetics* **63**, 283–289.

Vonesh, E. F., and Chinchilli, V. M. (1997). *Linear and nonlinear models for the analysis of repeated measurements*. Marcel Dekker, New York, NY.

Wackerly, D. D., and Mendenhall, W., III. (1996). *Mathematical statistics with applications*. Duxbury Press, New York, NY.

Wagenmakers, E-J, Farrell, S., and Ratcliff, R. (2004). Naïve nonparametric bootstrap model weights are biased. *Biometrics* **60**, 281–283.

Wallace, C. S. (2004). *Statistical and inductive inference by minimum message length*. Springer, New York, NY.

Weakliem, D. L. (Ed.) (2004). Model selection. *Sociological Methods and Research*, **33**, 167–304.

Wedderburn, R. W. M. (1974). Quasi-likelihood functions, generalized linear models, and the Gauss-Newton method. *Biometrika* **61**, 439–447.

Wel, J. (1975). Least squares fitting of an elephant. *Chemtech* Feb. 128–129.

White, G. C., and Burnham, K. P. (1999). Program MARK: Survival estimation from populations of marked animals. *Bird Study* **46**, 120–138.

White, G. C., and Lubow, B. C. (2002). Fitting population models to multiple sources of observed data. *Journal of Wildlife Management* **66**, 300–309.

White, H. (1994). *Estimation, inference and specification analysis*. Cambridge University Press, Cambridge, UK. pp. 380.

Williams, B. K., Nichols, J. D., and Conroy, M. J. (2002). *Analysis and management of animal populations*. Academic Press, New York, NY.

Woods, H., Steinour, H. H., and Starke, H. R. (1932). Effect of composition of Portland cement on heat evolved during hardening. *Industrial and Engineering Chemistry* **24**, 1207–1214.

Young, L. J., and Young, J. H. (1998). *Statistical ecology*. Kluwer Academic Publishers, London, UK.

Zellner, A., Keuzenkamp, H. A., and McAleer, M. (2001). *Simplicity, inference and modelling: Keeping it sophisticatedly simple*. Cambridge University Press, Cambridge, UK.

Zhang, P. (1992). Inferences after variable selection in linear regression models. *Biometrika* **79**, 741–746.

Zucchini, W. (2000). An introduction to model selection. *Journal of Mathematical Psychology* **44**, 41–61.

Zuur, A. F., Leno, E. N., and Smith, G. M. (2007). *Analyzing ecological data*. Springer, New York, NY.

# Index

AICc, 60–64, 66, 68, 72, 84, 86, 93, 96,
  113, 119
AIC values, 60, 63, 65, 66, 67, 84, 85
Akaike's information criterion (AIC), 55–60,
  60, 61, 62, 64, 65, 68, 70, 144
"All possible models," application, 140
ANOVA model, 8
Approximating model, 56, 57, 59
*A priori* hypotheses, 6, 9, 16, 78, 96, 116, 118
Asymptotic bias correction, 58
Asymptotic sampling distributions, 64
Averaging detection probability, 115

Bayesian approach, 88, 102, 110, 121,
  135, 158, 159
Bayesian information criterion (BIC),
  160–162
Bayesian Markov Chain Monte Carlo
  method, 135
Bayesian methods, 29
Beak lengths, modeling of, 37–41
Bias-corrected log-likelihood, 57–58
Bias correction term, 57, 61
Bias estimation, in statistical expectations,
  156
Bimodality, 37
Binomial coefficient, 155
Binomial distribution, 53
Binomial likelihood function, 36
Binomial model, 41
Boltzmann's entropy, 55, 144
Bootstrap in sampling variance
  estimation, 129
Bovine tuberculosis, 4, 69

Cement hardening data, 66–69, 112–115,
  119, 123
Chi-square distribution, 58
Chi-square test, 43
*Cinclus cinclus. See* European dippers
Coding theory, 55
Coin flip protocol, 136, 137
Conditional variance, 111, 113
Confirmatory studies, 8–10
Conflict resolution, information-theoretic
  methods usage, 135, 136
Conflict resolution protocol, 137–138
Constant hazard functions, 35, 36
Convenience sampling, 20
Correlation coefficient, 34
Count data overdispersion, 126
Covariance matrix, 25
Critical thinking, 10, 12, 15, 17, 37, 47,
  49, 73, 77, 78, 158, 164
Cross validation, 72
Cutting edge, in empirical science, 146

Data analysis, 8, 11, 15, 19–24, 59, 62, 63,
  98, 112, 118, 144
Data based model selection, 115
Data collection, analysis, 9, 12, 21
Data collection methods, 143
Data dredging, 45, 64
Data fitting, 25
Data information, 11, 20, 31, 60, 106
Data modeling, 21
Data optimization, 36
Data pattern, 46
Data *post hoc* estimation, 146

Data sampling, 14, 98, 116, 121
Data set, 22, 24
Degree of dependence and overdispersion
    parameter, 127
Degrees of freedom, 7, 43
Design based inference, 8
Deviance information criterion (DIC),
    135, 159
Deviance in likelihood-based inference,
    153
Discrete random variable, expectation, 155
Discriminant function analysis, 15, 24
Disease transmission, 5
Dominant variable in model selection
    bias, 129

Effect size, 7
Elephant seals, data overdispersion,
    128–129
Empirical science, 25, 26, 32, 33
Empirical support, 47
Entropy and information theory, 54
Entropy maximization principle, 55
Enzyme inhibition, 44
Enzyme kinetics, modeling of, 44–45
Error distribution, 64, 72
European dippers, 46
Evidence, 2–4, 8, 9, 11, 12–13, 14–16, 21,
    25, 26, 39, 41, 45, 47, 63–65, 78,
    83, 85, 87–100, 118, 120, 121, 133,
    135, 137–141, 143–146
Evidence ratio, 2, 89, 90, 92, 97, 99, 118,
    145
Experimental data, 4
Experimental design, 7, 19, 20

Fast learning strategy, 21
Ferrets, Bovine TB transmission in, 4–6,
    23, 35, 69, 93–94, 121
Finch bill lengths, unimodality of, 38
Fisher's likelihood theory, 55, 75
Fitted model, 56
Fixed effects models, 134
Flather's models, 95, 106, 110
Flood effect, on European Dippers, 95–98,
    109
Flour beetles, dose response modeling,
    41, 42
Force of infection, 4
Fourier series, 25, 121
Freedman's paradox, 16, 132, 145, 165
Full reality, 25, 27–29, 53, 54, 57–60, 62,
    66, 70, 75, 84, 85, 88, 99

Gamma distribution, 37, 38, 39, 53, 64
Gender effect, 36
*Geospiza fortis*, 37
Geostatistical modeling, 71
Goodness-of-fit (GOF), 43, 44, 87, 99
Goodness-of-fit test for data overdisperion,
    128, 146

**h**-likelihood, 135, 159
Hierarchical models. *See* Random effects
    models
Human medical research, 8

*iid* assumptions in statistics, 126, 127,
    147, 148
Independence, deficiency in count data, 126
Independent and identically distributed
    (*iid*), 147
Inductive inferences, 6, 20, 21, 98
Inference methods in life sciences,
    145, 146
Information loss, 66
Information-theoretic approaches, hen
    clam experiment, 138–140
Information-theoretic approaches,
    levels, 146
Information-theoretic approach (s), 11, 47,
    55, 64, 72–73, 95–96, 102, 110
Information-theoretic methods, usage in
    conflict resolution, 135, 136
Information theory, 52, 54, 74

**K**ullback-Leibler (K-L) information,
    52–56, 58, 59, 66, 70, 79, 84, 87,
    88, 95, 101, 121, 135, 144, 146

Least squares, 29, 34
Least squares estimates, 68
Least squares methods, 107, 110
Least squares regression, 61
Life sciences, 12, 72, 106
Likelihood-based estimation, 72
Likelihood data function, 41
Likelihood function, estimation, 148, 153
Likelihood methods, problems, 154
Likelihood model, 86–87
Likelihood ratio tests, 58, 153
Likelihood theory, uses, 149, 151, 152
Linearity, 25
Linear/logistic regression in model
    selection bias, 129
Linear models, 42
Linear regression models, 28

*Llikmalc*, in hen clam experiment, 138, 139
Logistic regression, 43
Log-likelihood function, 57, 61, 64, 65, 69, 72, 95, 117, 118, 150–152, 154
Log-likelihood of overdispersion parameter, 127, 128

Markov Chain Monte Carlo (MCMC), 158
MARK software, 123
Mathematical model, 21, 25, 72, 88
Mathematical modeling in empirical science, 143
Mathematical statistics, 54, 58
Matrix trace, 70
Maximum likelihood estimate (MLE), 11, 29, 34, 41, 42, 43, 45, 64, 66, 67, 68, 86, 96, 98, 107, 109, 110, 149, 153, 155, 156
Maximum likelihood model, 55, 57
Maximum log-likelihood function, 52, 56, 57, 61, 69, 72, 95, 135, 144
Mean squared errors, 135
Michaelis-Menten model, 44
Missing data, 65
Mixture distribution, 38, 39, 154
Model averaging, 106–110, 112, 113, 115, 131, 132
Model based inference, 9, 15, 21, 25, 26, 88, 120, 162–163
Model *g*, 53, 58, 60, 93, 111
Modeling hypotheses, 35, 36
Model parameters, 27, 56, 86, 99, 109
Model probabilities, 87–93, 95–99, 101, 106–109, 112, 113, 115, 116, 118, 119, 122, 131
Model probabilities $(w_i)$ estimation, 134
Model redundancy, 133, 134
Model selection, 9, 10, 59, 62, 95, 97, 106, 109, 110, 115, 116, 118, 119
Model selection bias, 129, 130
Model selection bias and applied data analysis, 130
Model selection bias problems, solutions, 130–132
Model selection bias reduction, model averaging in, 131
Model selection criteria in multivariate analysis, 133
Model selection uncertainty, 62, 63, 70, 84, 95, 109–115
Monte Carlo simulation, 23, 27
MSEs. *See* Mean squared errors

Multimodel inference, 99, 106, 108, 114, 115, 120
Multinomial distribution, 53
Multiple working hypotheses, 3–4, 9
Multivariate AICc, 133
*Mycobacterium bovis* infection, data of, 23, 24

Nested models, 47, 112
Nested *vs.* nonnested models, 63
Nonlinear regressions, 94
Normal distribution, 53
Null hypothesis testing, 3, 5, 7, 11–13, 46–47, 62, 64, 68, 90, 158, 164
Null model, 47
Numerical methods, 72

Objective bayesian approaches, 11
Observational studies, 10, 11, 15, 35, 44
Occam's razor, 31, 146
Overdispersed count data, 126
Overdispersion *(c)* parameter, in biological data, 127
Overdispersion in count data, coping with, 127, 128
Overdispersion in data on elephant seals, 128, 129
Overdispersion parameter estimation. *See* Variance inflation factor estimation
Overdispersion parameter usages, 128
Overfitting, 25, 30–33, 59, 62, 71, 81

Parameter count K, 71
Parameter estimation, 5, 23, 29–30, 53, 55, 56, 59, 63, 65, 71, 72, 86, 94, 107, 109–112, 114, 118
Parameter heterogeneity in overdispersed count data, 126
Parsimonious model, inference, 24
Parsimony principle, 30, 59, 70, 88, 101
*Plasmodium falcipraum* infection, 63, 71
Plausible hypotheses, 11
Plausible science hypotheses $(H_i)$, 3, 142
Poisson distribution, 53
Portland cement hardening data, 13, 22–23, 91, 112–114, 119
*Post hoc,* 64, 96–98, 118
*Post hoc* data analysis, 8, 9, 22, 45
Prediction theory, 108, 109
Predictive likelihood, 56, 68, 72
Predictive mean squared errors (PMSE), 59, 68, 162

Predictor variables, 119, 120
Pretending variable, 65, 117
Principal components analysis, 37
Principle of parsimony, 146
Probabilistic sampling, 7, 20, 23
Probabilistic theory, 19, 23
Probability and likelihood, comparison, 148, 149
Probability distribution, 53
P-values, 5, 7, 12

QAICc usage in overdispersed data, 128
Quasi-experiment, 7, 15
Quasi-likelihood concept, 128

Random effects models, 134, 135
Random sampling, 20, 23
Recapture probability $(p)$, 47
Regression analysis, 61, 94
Regression models, 107, 120
Relative ranking of importance, 119, 123
Residual sum of squares (RSS), 61
Response curve, dose, 41
Response variable, 108, 113, 119
Reversible jump Markov Chain Monte Carlo approach (RJMCMC), 159
$R^2$, likelihood version, 154

Sampling protocol, 19, 20, 21
Sampling variance, 110, 111
Schwarz' BIC, 160
Science and technology, 144
Science hypotheses, 11, 12, 25, 26, 62, 63, 64, 67, 69, 71, 72, 84, 85, 87, 93, 94, 98, 99, 107, 121
Second-order bias correction, AICc, 60, 65
Shannon entropy, 78
Size estimation and stratification, 7, 21
Squirrel occupancy, 116
Standard error (se), 7, 30, 67, 113, 114, 117, 118
Statistical expectations, types, 155, 156

Statistical inference, 12, 29
Statistical methods, 29
Statistical science, 6, 8, 11, 12, 23, 26, 32, 106
Statistical theory, 11, 54, 98
Stepwise regression, 33
Stochastic biological processes, 62
Stochastic complexity, 25, 62
Stochastic component, cement hardening, 23
Stochastic model, 25, 87

Tail probability, 13
Takeuchi's information criterion (TIC), 57, 70
Tapering effect sizes, 33
Threshold effects, asymptotes, 25
Trace operator, 70
Trade-offs, 30–32, 59, 101
Traditional null hypothesis testing, 46–47
Treatment *vs.* control groups, 7
*Tribolium confusu*, 41
True distribution $f$, 54
True models, 27–28

Uncertainty estimation, 87, 90, 95, 109–111, 113
Unconditional variance, 110–112, 114, 115
Underfitting, 30, 32, 59
Unimodal, 37–39, 154

Variance, 7
Variance component, 111, 113
Variance component models. *See* Random effects models
Variance-covariance matrix, 52, 128, 149, 152
Variance inflation factor estimation, 127
Variance, residual, 61, 66

Weibull errors, 25